WHO WERE

THE REAL BUFFALO SOLDIERS?

BLACK DEFENDERS OF AMERICA

*To Marilyn
Best wishes*

WHO WERE THE REAL BUFFALO SOLDIERS?

BLACK DEFENDERS OF AMERICA

Robert Ewell Greene

April 19, 2014

R.E. Greene, Publisher
Fort Washington, Maryland, 1994

Copyright © 1994 by Robert Ewell Greene
ALL RIGHTS RESERVED
Published in 1994 by R.E. Greene Publisher
Fort Washington, Maryland

Library of Congress Catalog Card No. 94-078121

ISBN 0-945733-12-7

To those "Real Buffalo Soldiers" of the Indian Wars, Spanish American War, Boxer Rebellion, Philippine Insurrection, Punitive Expedition, Mexico and their successors through the years who have so eloquently carried on the traditions of a Buffalo Soldier.

and

To the memory of my late paternal grandmother, Cecelia Spears Greene, whose genetic diversity of African and Seminole Indian in a spiritual sense has given me some understanding how two minorities, the American black and Native Indian did confront each other on the battlefield. She would probably say that was their way of survival in a white majority dominated society in that day and time. That was the way they were.

ILLUSTRATION CREDITS

I am always indebted to my sister, artist Ruth Greene Richardson, for her support of the cover paintings. However, I am most grateful for the creative cover painting that is beyond any doubt indicative of her successful years of painting. Time has not diminished your continual superior creative works of art, my dear loving sister Ruth.

ACKNOWLEDGEMENTS

I wish to express my sincere gratitude and appreciation to Janice Wood Hunter for her caring, patience, understanding and most efficient outstanding administrative assistance in the preparation of this book. Her performance has been exemplary.

I sincerely thank Viola F. Smith for her superb scholarly expertise in the grammatical editing and proof reading of this book.

I wish to thank a very knowledgeable friend whose vast experiences in the U.S. Military and his scholarly abilities as a Military historian, Lieutenant Colonel Major Clark, AUS Retired. I thank you again for your advice, materials and assistance you have provided during the preparation of this book.

I wish to express my sincere appreciation to First Sergeant Mark Matthews "A Real Buffalo Soldier" and his daughter Mary Watson for providing me an opportunity to interview her father.

I also wish to express my sincere gratitude to the following people for their contributions in making this manuscript a reality.

> Rev. Edward Calbert
> Dolores Camack
> Major Clark
> Bill Grant
> David A.F. Greene
> Janice Wood Hunter
> Kimberly Cherie Hunter
> Charles Johnson Jr.
> Tom Mackey
> Mark Matthews
> Viola Smith
> Mary Watson

My appreciation is also hereby expressed to the wonderful persons who have assisted me and may have been inadvertently excluded.

R.E.G.

CONTENTS

AUTHOR'S PREFACE

ACKNOWLEDGEMENTS

INTRODUCTION

CHAPTER 1	NINTH U.S. CAVALRY REGIMENT	3
CHAPTER 2	TENTH U.S. CAVALRY REGIMENT	23
CHAPTER 3	TWENTY-FOURTH INFANTRY REGIMENT	49
CHAPTER 4	TWENTY-FIFTH INFANTRY REGIMENT	57
CHAPTER 5	SEMINOLE NEGRO INDIAN SCOUTS	69
CHAPTER 6	PLAINS INDIANS	75
CHAPTER 7	INDIAN WARS	83
CHAPTER 8	SPANISH AMERICAN WAR	91
CHAPTER 9	BOXER REBELLION AND PHILIPPINE INSURRECTION	99
CHAPTER 10	PUNITIVE EXPEDITION, MEXICO	105
CHAPTER 11	BIOGRAPHICAL SKETCHES	113
CHAPTER 12	SAGA OF MARK MATTHEWS	189
CHAPTER 13	BUFFALO SOLDIERS - A COMMENTARY	237
CHAPTER 14	A PICTORIAL REFERENCE	249

APPENDICES		363
I	THEY DID NOT TELL ME	365
II	COLONEL CHARLES YOUNG	370
III	HENRY OSSIAN FLIPPER	388

BIBLIOGRAPHY 391

INDEX 401

Author's Preface

I sincerely believe that the time is now and has always been to reeducate and relate the true experiences of those Real Buffalo Soldiers - Black Defenders of America - especially when a polarized society, sometimes in the educational circles, does not present the presence of Africans Americans in some episodes of the American experience, especially in the conquest of the Far Western Frontier.

It is unfortunate that even in this day and time there is still a tendency for the mass news media to use radio, television, newspapers and magazines to minimize the positive laudatory achievements of African Americans in our multiracial society.

We as people of color need a more positive and brighter portrayal of our attributes and abilities as a people and of above all as African Americans. We need this to counteract if I may say the continual depiction of black people through the daily headlines and lead stories as the perpetrators, victims of drug abuse, criminal behavior, poverty, unemployed, welfare characters and the bottom victim of educational testings and upward mobility trends.

The presentation of these true facts about those "Real Buffalo Soldiers" can serve as a real image model for young African Americans and for non black people to be presented with a different perception of people of color in America.

It is my intent to continue to research true and factual history about the African American's presence and the part they played in the building and defending of these United States of America.

Robert Ewell Greene
August 4, 1994

INTRODUCTION

Introduction

This manuscript has been written to examine in more detail the significant events in the careers of those "Real Buffalo Soldiers" of yesteryears. It is also the intent of this book to address the soldier's families and how they performed heroically without the constant leadership of their white officers.

I have utilized primary sources from archival materials and my personal collections and library. A thorough literature search of unpublished and published sources were conducted. I used the oral tradition method during my interviews with a Real Buffalo Soldier, First Sergeant Mark Matthews who celebrated his 100th birthday on August 7, 1994. An extensive use of photographs are included to balance out the omissions which, by design or otherwise, have prevented actual portrayals in study after study, and volume after volume of the Buffalo Soldiers through the years.

I simply believe that this manuscript will reeducate some blacks, whites, Hispanics, Asians and foreign visitors that black people were present in other aspects, episodes and events of American History during the period, 1866-1943 than just their participation in music, comedy, song and dance, dribbling a basketball, hitting a ball and throwing a football. I often tell my college students that I do not make this statement in an offensive manner because my father was a physical education director, and participated in many sports since the age of ten years. I have his medals from his elementary year participation to his college days. His love for sports did not preclude him from obtaining a law degree. I have a second cousin who is a singer but she also pursued studies on the college level. There were many Buffalo Soldiers who served in the Military and then attained upward mobility trends in other professions.

The diverse biographical sketches in this book will indicate how many of the Buffalo Soldiers possessed numerous talents. I thought it was most interesting that the Buffalo Soldiers of the Tenth U.S. Cavalry enjoyed polo and fox hunts (a 300 year old English tradition). They also participated in pistol, rifle and horse show competitions. They enjoyed sports, had bands and musical shows, but they were not the dominating factors of their lifestyles as some writers have stereotyped them over the years.

It has been my primary goal to have this book portray to the reader beyond any doubt that those African Americans, people of color, affectionately known as Buffalo Soldiers served their country with fidelity, bravery and courage during the period 1866-1943. I sincerely trust that this book will illuminate the true meaning and most illustrious significance of the Fort Leavenworth Buffalo Soldier Memorial and the Postal Commemorative stamp of the Buffalo Soldier.

CHAPTER 1

NINTH U.S. CAVALRY REGIMENT

Ninth U.S. Cavalry Regiment

The Ninth Cavalry Regiment was constituted in the regular army on July 28, 1866. The regiment's organization commenced at Greenville, Louisiana on September 21, 1866. Units of the Ninth Cavalry Regiment received campaign participation credit for successful performance in the Indian Wars, involving the Comanches, Utes and skirmishes or battles at Pine Ridge, New Mexico, 1877, 1878, 1879, 1880, 1881 and Montana 1887, War with Spain, Santiago and the Philippine Insurrection.

Ninth   The Heraldic Items of the cavalry were: "Coat of Arms"

Shield: Or, on a pile Azure in Chief a sun of eight points of rays between three five pointed mullets two and one of the field, in base over all the blockhouse of San Juan Hill, Santiago, Cuba, proper.

Crest: On a wreath of the colors a horseshoe with nail holes heels down argent winged purpure depressed by two arrows in saltire sable armed and flitted gules.

Motto: We can, We will

Symbolism: The field is yellow for the cavalry. The blue triangle with the sun and three five pointed stars are from the old flag of the Philippine Insurrection with a change of color. The three stars also represent three tours of duty in the islands. The blockhouse is the old pride of the regiment, a representation of the actual one taken at San Juan Hill in 1898. The wedge is blue and recalls the fact that the Ninth split the Spanish line at Santiago with the capture of the blockhouse when it charged dismounted as infantry. The crest is the well known scotch device signifying the alertness of the mounted man, and the arrows are for the Indian campaigns of the regiment.

Distinctive Insignia

The distinctive insignia is an Indian in breech clout and War bonnet, mounted on a galloping pony, brandishing a rifle in his right hand and holding a single rein on his left hand all in gold, displayed upon a five-bastioned fort in blue edged with gold.

The five bastioned fort was the badge of the Fifth Army Corps in Cuba, of which the Ninth was a part. The yellow outline is for the Cavalry, and the blue for active service in the Spanish American War. The mounted Indian represents the Indian Campaign of the regiment."

The Ninth Cavalry Regiment, an outstanding regiment of the "Real Buffalo Soldiers" and in later years their successors, was formally inactivated on March 7, 1944 in North Africa. It is to be noted that the Fourth Squadron Ninth Cavalry was demobilized December 1, 1921 in the Philippine Islands and Troop E was disbanded October 20, 1950.

Where Were They?

The "Real Buffalo Soldiers" of the Ninth Cavalry Regiment were there in the "Winning of the West", Spanish American War, Philippine Insurrection, Boxer Rebellion, and the deployment of military troops in civilian strikes, unrest and protests. They were there and unfortunately the following historical facts are still missing from many educational textbooks on levels from elementary to college.

Stage Coach Escorts

1867 One hundred Mescalero Apaches attacked a stage coach eastbound from El Paso, Texas and killed the escort soldier, Private Nathan Johnson and another soldier was wounded. Company F, Ninth Cavalry Regiment was able to cause the Indian warriors to retreat near Eagle Springs, Texas.

Mail Escorts

1867 There were some members of Company D, Corporal Emanuel Wright and private E.T. Jones who were killed when ambushed by a band of Kickapoo warriors. Wright and Jones were performing one of their major duties, escorting mail from Camp Hudson to Fort Stockton.

Pursuit of Apache Indian warriors

September 18, 1868 Colonel Merritt sent Lieutenant Patrick Cusack and 60 Buffalo Soldiers of Ninth Cavalry Regiment along with 10 civilians to pursue 200 Indian warriors who had attacked a train near Fort Stockton, Texas. The warriors went into the Santiago Mountains of Big Bend. Lt Cusack charged the Indians and were successful in a victory. Twenty-five Apache Indian warriors were killed and some were wounded. The troopers rescued two Mexican children. After their successful encounters with the Indians the soldiers returned to Fort Davis.

Chasing Kiowas and Commanche Indians

1869 Buffalo Soldiers of Companies B,E,F and M pursued some Kiowa and Commanche Indian warriors near Fort McKavett. They lost the Indian's trail and returned to Fort Concho.

October 10, 1869 Later, the Buffalo Soldiers were able to confront some Indians in the area of Salt Fork of Brazos. They had traveled 600 miles in 42 days. The Fourth U.S. Cavalry and the Tonkawa Indian Scouts also participated in the battle.

An Expedition Against Apaches

1869-1870 Colonel Edward Hatch, Ninth U.S. Cavalry Regiment dispatched three expeditions against the Apache warriors in the Guadalupe Mountains in 1869-1870. On January 20, 1870 Captain F.S. Dodge and his Buffalo Soldiers killed 25 Apaches warriors and captured some stock.

A Skirmish With Indian warriors

1870 A small detachment of the Ninth Cavalry Regiment was searching for cattle rustlers in January, 1870. The trooper observed some Indians and the detachment's non Commissioned Officer (NCO), Sergeant Edward Troutman confronted the Indians in a skirmish. One Indian was killed and two Ninth Troopers were killed.

Texas Racism

1870 In 1870, some white settlers shot and killed a Ninth Cavalry Regiment soldier, Boston Henry at an area near Fort McKavett, Texas. When two Ninth Cavalry soldiers were sent to apprehend the alleged murderers, they were killed. The suspected men were found not guilty in a Texas court.

Pursuit of Apache Warriors

1871 When a raiding party of 15 Apache Indians stole 41 army mules and 13 horses at Bariela Springs, Texas, Colonel Shafter responded to the incident. Shafter sent 63 mounted soldiers of the Ninth Cavalry Regiment to follow the Indian's trail. The Buffalo Soldiers marched for two weeks in areas near the south western corner of New Mexico. They traveled "the rolling dunes of the Monahani Sands." The soldiers had marched 70 miles without water before they located an Indian village of 200 people. The cavalry soldiers returned on July 9, 1871 to Fort Davis, Texas, after 22 days in the field.

Ninth Cavalry Support Texas Rangers

1877 On December 15, 1877, Governor Hubbard of Texas requested federal assistance in keeping law or order in El Paso, Texas during the "Salt War". Mexican and Americans were in disagreement over salt mines in the Rio Grande area. Colonel Hatch sent some Buffalo Soldiers under the command of Lt. Rucker to assist the Texas Rangers. Nine companies of the regiment were sent to provide a show of force. The soldiers marched to San Elizario. Eventually, the rioters began to disburse. This was another example of African American troops being used to settle disputes between whites and Mexicans.

Utes Indians Unrest

March 1878 There was some unrest among the Colorado Utes Indians in March of 1878. Indian Agent Weaver feared there could be some violence and requested assistance from the Ninth Cavalry Regiment. The Utes were living along the White River near the Mexican border. Major Morrow and companies D,G,I and K marched to the La Plata River. Later Colonel Hatch and Captain Kimble attempted to have a meeting with the Southern Ute War Chief, Ignacio. Agent Weaver had plans for the Utes to adopt an agricultural life style, working as farmers. The Utes were not interested in Weaver's plans.

Ninth Cavalry Search Party

June 1878 Indian warriors were raiding in the Guadaloupe Mountains and Ninth Cavalry search parties were sent out to locate the Indians. Captain Carroll left Fort Stanton with Buffalo Soldiers from companies F and H, and some Navaho scouts.

The Lincoln County War and Ninth Cavalry

Lincoln County, New Mexico is located in southeastern New Mexico. In 1878, the population was around 1,800 citizens. There were numerous disputes between business men, cattle ranchers, law enforcement officers and outlaws. Henry Atrum later known as Bill Bonney and better known as Billy the Kid was involved in some of the disputes and what was popularly known as the "Lincoln County War". The two opposing factions were the county sheriff supported a judge, and district attorney and a posse and the other faction a group known as the Regulators who had supported by a special constable and a Justice of Peace. The two groups were engaged in some illegal transactions. Their differences became violent at times and some individuals were murdered. Somehow, the sheriff with assistance from the governor was able to use the military to maintain peace and order and present "a show of force". The nearest military fort was Fort Stanton where the Ninth U.S. Cavalry Regiment was stationed. The commanding officers of the Ninth Cavalry Regiment deployed men of the regiment in Lincoln County on search occasions and also in the pursuit of Billy the Kid.

On February 18, 1878, Captain George Purington and a detachment of the Ninth Cavalry arrived in the town of Lincoln to support a lawyer, named Alexander McSween. A request was made for the Ninth Cavalry to provide assistance in the town on April 29, 1878. Lt. George Smith led a detachment of 20 Buffalo Soldiers of the Ninth Cavalry.

When Billy the Kid and other outlaws left the town, a detachment of Ninth Cavalry troopers departed from Fort Stanton to search for them, on June 19, 1878.

On July 19, 1878, a detachment of the Ninth Cavalry and some white soldiers of the 15th Infantry Regiment were sent to the town, to assist Sheriff Peppin.

The commanding officer of the Ninth Cavalry and companies F and H marched into Lincoln County in July 1878 with a howitzer, gatling gun, and rations for three days. A request was made to the justice of peace to obtain a warrant for the arrest of Billy the Kid.

On October 2, 1878, Captain Carroll arrived in town with 20 Buffalo Soldiers to provide a show of force at Bartlett's Ranch.

In 1879, the town's sheriff requested assistance from the Ninth Cavalry Regiment. Lt. Kimball and a detachment of the Ninth Cavalry went into town.

The Ninth U.S. Cavalry's Buffalo Soldiers were there in 1878-1879 maintaining law and order when requested in the civilian community and also pursued Billy the Kid and his fellow outlaws.

Theft of Cattle by Indians

September 1879 A herd of cattle belonging to Company E, Ninth Cavalry Regiment were stolen by a band of Indian warriors. This occurred near Ojo Caliente.

Indians Attack 9th Cavalry Troopers

September 1879 While pursuing Victoria, Captain Dawson and his Ninth Cavalry Buffalo Soldiers were attacked by some Indians near Las Animas River. He received assistance from Captain Beyer and his troopers. The Indians continued to fight and eventually the Ninth Cavalry were forced to return to their forts. There were some casualties and a loss of 32 horses.

A Skirmish with Indians

September 29
1879 Major Morrow was leading a contingent of 200 men near Ojo Caliente. He was engaged with Indian warriors in a skirmish for two days. His men were without water for 3 days and also rations Their horses were exhausted.

A Skirmish and Pursuit of Indians

October 27
1879 On October 27, 1879, Major Morrow and his Ninth Cavalry Troopers were successful in a skirmish with some of Victoria's Indian warriors. The fight occurred 12 miles from the Corralita River, Mexico. he pursued the Indians with his Buffalo Soldiers and later they returned to Fort Bayard, New Mexico on November 3, 1879.

Pursuit of Victoria

January 2
1880 Victoria and his warriors were raiding and murdering in southern New Mexico in January 1880. The military command was concerned about Victoria and his men using the Mescalero Agency at Fort Stanton as a supply base for their rations and recruitment of additional warriors. The commanding officer of the 9th Cavalry, Colonel Hatch received orders to form a task force to continue the pursuit of Victoria. Hatch organized a force of 400 cavalrymen to include 60 infantrymen and 75 Indian scouts.

Skirmish at Tularosa

April 16,
1880 Major Morrow and his Ninth Cavalry Troopers were able to capture some Indians and continue their pursuit of some forty Indian warriors who had escaped toward Dog Canyon. The skirmish occurred at Tularosa. Three Indians were killed and 25 head of stock captured.

A Skirmish with Indians

August 3, 1880 A patrol under Corporal Asa Weaver, Ninth Cavalry Regiment was involved in a 15 mile running fight near the Alamo. The horse of Trooper Willie Tucker buckled and ran into some charging Indians.

The Pursuit of Indian Leader Nana

July 1881 The Ninth Cavalry Buffalo Soldiers pursued Nana and his Indian warriors in July 1881. They were traveling along the trails of the White Sands and San Andres. Captain Guilfoyle and 20 men of Company L along with Captain Parker and his Company K soldiers had chased Nana toward Sabinal.

The Ninth Cavalry and the Oklahoma Boomers

1879-1889 A proclamation was issued by President Rutherford B. Hayes on April 26, 1879 prohibiting citizens to enter the territory reserved for Indians. There were some people who tried to enter the Indian territory. (present day Oklahoma). These individuals were called "Boomers". It was necessary at times to use federal troops to prevent those potential settlers from occupying the land. The Ninth U.S. Cavalry Regiment was used to escort the Boomers out of the territory.

In April 1879, a group of white settlers under the leadership of Colonel Charles Carpenter traveled through the areas of Wichita and Coffeyville, Kansas. Later, Carpenter was succeeded by Captain David L. Payne known as Oklahoma Payne. It was believed that Payne had some financial assistance from the Atlantic and Pacific Railroad. Payne tried on several occasion to move into the Oklahoma Territory. Captain Payne decided to make a raid into the Oklahoma Territory. He had been warned to keep out by a Fort Worth, Texas Court. In 1882, Payne and 22 men crossed the border and entered into the Oklahoma Territory. In August, 1882, the Buffalo Soldiers of the Ninth and Tenth Cavalry were sent to arrest Payne and his men and escort them outside of the territory.

The Boomers were persistent to settle in the new territory. A Captain Couch succeeded Payne. He lead a group of 400 men into the Oklahoma territory to colonize some area of the territory. Colonel Hatch, commanding officer of the Ninth Cavalry Regiment dispatched Lieutenant Day and his Buffalo Soldiers to Couch's encampment at Stillwater, Oklahoma. The Boomers and Captain Couch were escorted by the Ninth Cavalry Regiment from the territory.

The Ninth Cavalry troops consisted of Company F, I and G. The men faced some severe weather conditions at times. The territory was finally opened to the settlers in 1889. Some 100,000 settlers from the states of Kansas, Texas and Arkansas entered the territory on April 22, 1889. Oklahoma became a state in 1907. The Buffalo Soldiers of color were used to preserve the Indian's privacy and rights on their land until 1889. The soldiers were able to arrest and escort white settlers out of the territory. This is just an inference. In the 1920's, there was a serious race riot in Tulsa, Oklahoma between whites and blacks. Some of those whites and blacks could have been the original settlers or descendants of the settlers. Yes, Oklahoma was experiencing racial conflicts that could have involved the presence of the Indians, some early African American settlers and the white settlers whose possible goals were to dominate the political and economical climate of the new state. Also one must remember that Plessy v. Ferguson ruling was well enforced throughout America.

Ninth Cavalry Regiment at Wounded Knee

1890 The unfortunate massacre of Native Americans, Sioux Indians at Wounded Knee, South Dakota occurred in December, 1890 and will always be a sad note in the historical feats of the U.S. Military in the conquest of the Far Western frontier. Yes, the men of color, the Real Buffalo Soldiers were there. What was their role? They were there to assist the white soldiers of the Seventh U.S. Cavalry Regiment after their confrontation and killing of some innocent Indian people. The Seventh Cavalry was attacked by some Indian warriors on December 30, 1890 at a Catholic mission at Clay Creek. The Indians had encircled the white cavalry troopers. Approaching from the rear were the Buffalo Soldiers of the Ninth Cavalry led by a member of the 9th Cavalry, Corporal William O. Wilson. (see biographical sketch).

Ninth U.S. Cavalry Regiments' Battles and Engagements

1867-1900

DATE	PLACE
1867	
Dec. 5	Eagle Springs, near Fort Davis, Texas
Dec. 26	Fort Lancaster, Texas
1868	
Sept. 14	Near Horsehead Hills, Texas
1869	
Sept. 16	Brazos River, Texas
Oct. 18	Near Double Mountain & Clear Mountain Forks of The Brazos River, Texas
Dec. 25	Fort Concho, Texas
1870	
Jan. 20	Guadalupe Mountains, New Mexico
May 20	Near Fort McKavett, New Mexico
1877	
Jan. 24	Bocas Grande Mountains, New Mexico
Jan. 28	Bocas Grande Mountains, New Mexico
1878	
Aug. 5	Dog Canon, New Mexico
Sept. 4,28	Near Ojo Caliente, New Mexico
Sept. 18	Rio Las Animas, New Mexico
Sept. 27	Near Gonsuma Mountains, New Mexico
Sept. 28	Miembres, New Mexico
Sept. 29	Miembres Mountains, New Mexico
Sept. 30	Near Paraje, New Mexico
Jan. 12 & 17	Rio Perchos & San Mateo Mountains, New Mexico
Jan. 12 & 17	San Mateo Mountain, New Mexico
Jan. 12,13,17	Near Fort Bayard, New Mexico
Feb. 3	Miembres Mountains, New Mexico
Feb. 28	Alamo Canon, Sacramento Mountains, New Mexico
April 3,6,7	Miembrillo Canon, New Mexico
Sept. 1	Agua Chiquita, New Mexico
1881	
July 19	Near Arena Blanca, New Mexico
July 25	San Andreas Mountains, New Mexico
Aug. 3	Monaco Springs, New Mexico
Aug. 12	Near Sabinal, New Mexico
Aug. 13	Near Camp Canon, New Mexico

Aug. 16	Cushillo Negro, New Mexico
Aug. 16	Nogal Canon
Aug. 19	Near McEwer's Ranch
Aug. 19	Membres Mountains
1887	
Nov. 5	Crow Agency, Montana
1898	
July 3-17	San Juan Ridge, Cuba
1900	
Nov. 25,26,27	Jovellar, Philippine Islands

The Ninth Cavalry In The Philippines In 1922

The Ninth Cavalry Regiment was stationed at Camp Stotsenburg, Pampanze, Philippine Islands (P.I.) 1918-1922. The regiment's strength was 552 enlisted men, four cavalry officers, one chaplain, two field artillery officers, and ten enlisted medical detachments.

The regiment departed the P.I. on October 11, 1922 aboard the USS AT. Logan, destination the United States. Their route to the U.S. consisted of stops at Nagasaki, Japan, Honolulu, Hawaii, and final destination, San Francisco, California on November 11, 1922.

The return of the Ninth Cavalry regiment to the United States in 1922 was the beginning of a distribution of the regiment personnel to other units and headquarters. Personnel were reassigned as follows: "The Headquarters, Troop", Service Troop, 1st and 2nd Squadron headquarters and nuclear of personnel from each troop, a total of 204 men, to Fort Riley Kansas. One master sergeant, one staff sergeant, eight sergeants, seven corporals and forty-three privates to Fort Leavenworth, Kansas. The remaining personnel of the regiment 344 men were transferred to the Tenth U.S. Cavalry Regiment at Fort Huachuca, Arizona.

The Ninth U.S. Cavalry Regiment, Fort Riley, Kansas, 1939-1940

The entire regiment of the 9th Cavalry was stationed at Fort Riley, Kansas in 1929 and performed duties at the Cavalry School. In 1940, some members of the regiment were detailed to furnish assistance in Minnesota, North Carolina, and Oklahoma. On June 11, 1940, the 9th Cavalry was detailed to provide 8 scout cars with crew for use with an umpire group during principal maneuvers August 11-17 at Little Falls, Minnesota. The

16 The Real Buffalo Soldiers

detachment consisted of "9 scout cars, one truck, one motorcycle with driver and 41 enlisted men. The detachment arrived at Lastrik, Minnesota on August 6, 1940. The detachment operated for 1,000 set hours of continuous operations with less than 3 set hours of breakdown due to set failure or vehicle failure."

The Ninth Cavalry provided a cadre of "5 corporals on August 6, 1940 for the 76th (CACAA), Fort Bragg, N.C. On August 9, 1940, a cadre consisting of 3 First Sergeants, 1 technical sergeant, 13 sergeants, 2 acting first sergeant, 33 privates first class was sent there also".

The Ninth Cavalry and Biogenetic Diversity In The Philippines, In 1922

"What About The Mothers and Children"

The United States Military has stationed troops in many areas of the world. Sometimes these troops remain for a number of years. A problem that is often addressed by military and civilians of the host countries is the fraternization of servicemen with ladies of the respective countries. Often a relationship develops into marriage and/or the birth of children out of wedlock. This has occurred since the 1900's and the military has dealt with these situations in World Wars I and II, Korean War and the Vietnam War. There are many happy and long relations in marriage between military men of all services with foreign women from Asia and Europe. Their children in the majority of cases have adjusted into the American way of life as another American child of military or former military parents.

There is a problem that the U.S. Military has had to address in World War I and II, Korea and Vietnam Wars, especially World War II, with the coming home of African American soldiers from Europe with their French English and Italian brides.

After the wars or conflicts, the countries were left with some brides and/or girl friends who had children fathered by white and black servicemen. In some countries there were no problems in locating orphanages and adoption centers to accept the children of white Americans. However, that problem of skin color did exist when the miscegenated or mixed babies/children of African American fathers were present. A question that has often been asked, where are the mixed children of African American fathers and their mothers of Korea and Vietnam? Also how many children of African fathers actually boarded that mass airlift of young children from Vietnam at the end of the conflict. Today America has a serious problem of having babies of black parents being adopted by blacks and of course non blacks. Therefore, there would also exist a problem of adopting biogenetic diverse children from foreign mothers.

The problems of the disposition of children born to foreign mothers and African American did not originate in 1945. There was some concern as early as 1922. A concerned and thoughtful post commander developed a plan for the disposition of 207 Filipino women and 72 children of the 9th Cavalry and other "colored soldiers". The post commander of headquarters at Camp Stotsenburg, Pampanga, Philippine Islands (P.I.) submitted his plan and recommendations to the commanding general, Philippine Department, Manila (P.I.). The plan classified the wives and non wives and children into four categories:

"<u>First</u>. 37 Filipino women legally married to soldiers, having 56 children by them.

<u>Second</u>. 9 Filipino women not legally married to soldiers but living openly with them, and having 16 children by them.

<u>Third</u>. 95 Filipino women legally married to soldiers but having no children by them.

<u>Fourth</u>. 66 Filipino women and one Japanese not legally married to soldiers but living openly with them and having no children."

The Post Commander viewed the first category as presenting great difficulty. He believed the U.S. Military should not compel, encourage or allow these soldiers to abandon their lawful wives and children. He wrote, *"This should not occur merely for the purpose of conforming to a change of policy based on some minor economy or convenience to the government. On the other hand, it seem to be impracticable to take these 37 families to the United states. Transportation having been provided as far as San Francisco, very few soldiers would have the money to pay railroad fare to Fort Riley or Fort Huachuca, and even if they were taken on troop trains no quarters would be provided for them after arrival. Climatic conditions would not suit them, and if their husbands, at some future time, failed to reenlist or were otherwise discharged, these native women and children would be a charge upon the United States"*. The commander also addressed government policies. He said, *"It seems inconsistent with the policy of a generous government to discharge these men for its own convenience without making any provision, or allowing them any opportunity for securing proper employment. I cannot believe that if the facts were understood, the War Department or higher authority on account of color would compel the immediate discharge of high type, faithful and efficient soldiers of long service in such numbers as to embarrass the local labor market and impose an enormous hardship upon them without even giving them an opportunity to complete their current enlistment. Besides this, the insular government would probably object to such a procedure, as a number of such discharged soldiers are already open to the charge of vagrancy".*

The post commander saw the third category presented a similar difficulty but to a lesser degree. His reason was that 95 women without children that were easier to handle than the 37 women with children. The 9 women and 16 children of the class two presented a different problem, because these women could not be taken to the United States, unless the men married them and then they could be placed in class one. They would have to provide for them in the P.I.

The commander also believed that the only problem presented by the women in the category four was to get rid of them from the neighborhood, unless the men would marry them and place them in class three.

The commander had to address his views of the "so called paternal slave master when he said, "*We cannot assume that the heart of a colored man is any less sensitive to the destruction of his black family ties than that of a white man. No decent slave owner in the old south, would consent to the breaking up of families (the slave owner broke up families when he separated them by selling slaves from families), and the whole world was sickened by the sight of Germany's disregard of family ties in Belgium. It is inconceivable that these families should be broken up and abandoned merely to carry out some little unimportant detail of the policy, that can be changed by the stroke of a pen, along with hundreds of other changes being made every time a new man gets into a new position".*

The commander decided upon his solution. He said they were based upon the following principles:

"<u>First.</u>   *We must be governed by the ordering principle of humanity.*

<u>Second.</u>   *We must be fair in our treatment of soldiers who have rendered honorable and faithful service.*

Third.   Our plan must be acceptable to the government general in so far as it affects the civil community.

Fourth.   The efficiency of the army must not be sacrificed

Fifth.   The plan must be such as to minimize subsequent controversy.

The post commander solutions were:

"A.   Many of the women and children be sent to the United States as can be properly cared for and that the soldier responsibility for the rest of them be retained in the service in the P.I. until they can be otherwise disposed of.

B.   That a cable be sent to the War Department requesting information as to whether in the case of soldiers without necessary funds, the women and children in category one and the women of category three shall be sent to the United States (U.S.) with the troops and if so that proper arrangements be made for their transportation to the new stations and their accommodations after arrival.

C.   That a careful canvass should be conducted to determine those men who desire to be discharged and who are prepared to take care of their families in the P.I. These men to be so discharged. A canvass should also determine the native women who are willing to give up their husband and who consent to their husbands return to the U.S. without them. Affidavit to be secured from these women to this effect and the men so returned.

D.   That all men of categories one, two and three not disposed of a,b and c above, be retained in the military service in the Philippines and gradually discharged for the convenience of the government as rapidly as they can obtain suitable employment, or otherwise make their own arrangements for the proper disposition of their family ties. The following assignments are suggested for these men while being retained in the service. A certain number preferably the older non commissioned officers (NCO) could be retained as instructors in the new 26th Cavalry regiment. Of the balance some could be transferred to the medical department, some to the quartermaster corps and others formed into labor detachments. They could be used as teamsters, orderlies, messengers, mounted military police and on other necessary duty.

E. That the disposition of the 67 women in category four be held in abeyance until some policy is outlined for the dispositions of others. But that a careful investigation be made of each case with a view to obtaining a release of the man and to adjust all differences as to debts, ownership of property, etc. Then the soldier, could be returned to the United States."

The commander also stated that the whole question was more or less a local problem and the commanding general must expect to bear the burden of it and work out the details. Therefore the commander made a request for approval as submitted.

The Commanding General, Philippine Department, Fort William McKinley, Rizal, approved the plan. He stated that he had discussed the plan with the governor general who seemed satisfied with the plan.

There is evidence that some parts of the plan were executed because a memorandum to the commanding officer, Fort Riley Kansas from the office of the post commander, headquarters Camp Stotsenburg, P.I., dated October 8, 1922 was received. The subject of the memorandum was *"Relation of certain soldiers of the Ninth Cavalry to native women."*

There were listed on the memorandum the names of one first sergeant who was released from all obligations to a Filipino woman with no children. A staff sergeant was also released from obligations and the Filipino woman with no children would remain in the islands. Two privates were released from all obligations when the Filipino woman signed the affidavits. One of the privates was married and had no children and his wife decided to remain in the P.I. There were two privates who were married and their Filipino wives accompanied them to the United States.

A corporal's Filipino woman friend, not married and no children signed an affidavit refusing to release the soldier from obligations. A sergeant who was not married to a Filipino woman would remain in the P.I. with their one minor three month old child. The sergeant was unable to make an allotment for the Filipino woman because he had an allotment of fifty (50) dollars a month to a wife in the United States. The soldier agreed in writing to send ten (10) dollars a month until April 1924 for support of the child.

There are many inferences that can be made concerning the Ninth Cavalry soldiers and their Filipino wives and friends in 1922. The military viewed the situation as a problem. They were concerned about the welfare of the child and mother and also obligations that the soldiers could confront.

There was no discernible instances of racial factors involved in correspondence reviewed. However, one must be realistic in the view of the climate of racial segregation in America and the white majority in power enforced all facets of segregation where possible from a legal justification. Therefore, I infer that the problem was not a racial one. Because from a biological perspective, the military authorities were concerned with two peoples of similar skin color in many cases, simply "people of color". Of course, that can be debatable in America because unfortunately people are classified by sight and descriptions such as "hair, lips and skin color". I ask the question, why weren't some of the procedures used by the military in 1922 considered in relations to those brown babies in France, Germany, England, Korea and Vietnam. Whatever the answer may be, I deeply believe that many citizens of the Filipinos in 1922 probably would have said, a person can be accepted by their character, sincerity, morality and honesty and not be so concerned about one's skin color. I salute the concerned white post commander and his superiors who concurred with his recommendations because they were ahead of their time when concerned about the family and not just welfare and who pays for it.

CHAPTER 2

TENTH U.S. CAVALRY REGIMENT

CHAPTER 2
Tenth U.S. Cavalry Regiment

The 10th U.S. Cavalry Regiment was constituted in the regular army on July 28, 1866. The regiment's organization was commenced at Fort Leavenworth, Kansas on September 21, 1866. Units of the 10th Cavalry Regiment received "Campaign Participation Credit for successful performance in the *Indian Wars*, involving the Comanches and Apache's and Skirmishes or battles at New Mexico 1880 and Texas 1880. *War with Spain* Santiago and the *Mexican Expedition* Mexico 1916-1917.

The Heraldic Items of the 10th Cavalry are:

Badge Description: On a heraldic wreath or and sable, a buffalo statant proper. On a scroll of the second Fimbriated of the first the motto ready and forward of the like.

Symbolism: Black and gold have long been used as the regimental colors. The buffalo has like wise been the emblem of the regiment for many years, having its origin in the term *"Buffalo Soldiers"* applied by the Indians to the Negro Regiments.

Distinctive Insignias

The distinctive Insignias of the regiment were the badge and later a Regimental Coat of Arms.

Regimental Coat of Arms Description:

Per pale, dexter: paly of thirteen argent and gules, a chief azure, over all an Indian Chief's war bonnet afronte in chief and a tomahawk and stone ax in saltire, heads downward, in base all proper.

Sinister:
> Per fess, in chief quarterly, I and IV, gules a tower triple - towered or gated azure. II and III argent a lion rampant gules, crowned with a ducal cornet or; upon an oval in escutcheon azure, a fleur - delis or. In base sable, the katipunan device on its base, thereon the sun in its spendour, between three mullets, one and two, all or.

Crest: On a wreath, or and sable, an American Bison statant, proper. Beneath the shield, a riband sable, thereon "Tenth Cavalry" or.

The 10th Cavalry Regiment an outstanding regiment of the "Real Buffalo Soldiers" and in later years their success was formally inactivated on March 20, 1944 in North Africa.

Where Were They?

The "Real Buffalo Soldiers" of the Tenth Cavalry Regiment were present in the "Winning of the West", Spanish American War, Philippine Insurrection, Punitive Expedition to Mexico and the deployment of military troops in civilian strikes, unrests, and protests. They were there and unfortunately the following historical facts are still missing from many educational textbooks on levels from elementary to college.

Tenth U.S. Cavalry Regiment

A Historical Revisit

The Real Buffalo Soldiers" of the Tenth Cavalry Regiment were organized on July 28, 1866 at Fort Leavenworth, Kansas. Their first commanding officer was Colonel Benjamin H. Grierson. On August 6, 1867, regimental headquarters was moved to Fort Riley, Kansas. The last company formed was company M, called the "Calico". The regiment's first encounter with Indian warriors occurred on August 2, 1867,

1867 Company F of the regiment was given a detail assignment to patrol the Kansas and Pacific Railroad near Fort Hays, Kansas. Thirty-four enlisted men and one officer were attacked by 300 Indian warriors. The soldiers fought bravely until they were forced to retreat. A Captain G. A. Armes was wounded and their first casualty was Sergeant William Christy of Mercersburg, Pennsylvania. During extensive research on my book *Swamp Angeles a Biographical Study of the 54th Massachusetts, Regiment,*

I reviewed several pension files of the Christy brothers from Mercersburg, Pennsylvania. A Jacob Christy was born a slave of a white man and black woman. He was a slave of his illegitimate grandfather until he was 28 years old. In 1835, he married a Catherine Selver. She died on September 7, 1853. They were married by Reverend Robert Kennedy, of the Welsh Run, Presbyterian Church, Pa. Jacob and Catherine were the parents of five sons, Jacob, John, Joseph, Samuel and William, and two daughters Mary Jane and Elizabeth. When the Civil War started, the five brothers enlisted in the Union Army. Jacob, Joseph, Samuel and William all enlisted in Company I, 54th Massachusetts Regiment and were present at Fort Wagner and Olustee, Florida. John Christy had enlisted in Co. H, 45th Regiment United States Colored Troops (USCT) Tenth Corps. The sister Mary Jane Christy exchanged letters with her brothers during the Civil War. Several letters indicated that their brother William was wounded and possibly taken prisoner by the Confederate troops. In 1868, William Christy's father Jacob Christy applied for a "father's Pension". His claim stated that his son William Christy was born in Mercersburg, Pennsylvania in 1842 and that he enlisted on April 22, 1863 at Readville, Massachusetts and that "Christy was killed at Olustee, Florida on February 20, 1864." There was some speculation that he could have died as a prisoner of war. Jacob Christy's affidavit stated that he provided support to him prior to his enlistment. I made a visit to Mercersburg, Pennsylvania in the Spring of 1994 and I was not able to locate any descendants of the Christy family. However, I did see the grave site of one of the Christy brothers buried in the local cemetery and also a house in a nearby town where a niece of the brothers lived some years ago. I was able to ascertain that the Christy family did live in Mercersburg, Pennsylvania 150 years ago. During a literature search for this book, I learned that many sources list a name of William Christy killed in 1867 and was from Mercersburg, Pennsylvania. In consideration of the unusual or rarity of the name "Christy" for people of color and the designated geographical location of his home, also the fact that he would have been a young man at the age of 25 years old in 1867, it is a great possibility that the William Christy of the 54th Massachusetts was not killed at Olustee, Florida or died in a confederate prison. Some documents listed a William Christy from the 54th Massachusetts Regiment as missing in action. Therefore, I infer that the William Christy of the Tenth Cavalry Regiment who died during the Indian Campaigns in 1867 was the same William Christy from the 54th Massachusetts Regiment who probably did not die until 1867.

Tenth Cavalry Assist Seventh Cavalry

1867 In June 1867, some members of the Tenth Cavalry Regiment assisted the Seventh Cavalry Regiment at Fort Wallace, Kansas. Ms. Elizabeth Custer, wife of General George Custer who led the diasterous charge of the Seventh Cavalry at Little Big Horn wrote an account in her diary about the Tenth Cavalry. She stated that the Seventh Cavalry was engaged in a fight outside of Fort Wallace, Kansas, in June 1867. A Seventh Cavalry troop was involved in a counter charge by some Cheyenne warriors. Twelve members of the Tenth Cavalry were at the Fort to obtain some supplies. They assisted the Seventh Cavalry in fighting the Indians outside of the gate at Fort Wallace, Kansas.

Buffalo Soldiers Rescue Railroad Crews

1867 It was reported that on June 29, 1867, a unit of the Tenth Cavalry Regiment came to the rescue of some railroad construction crews near Fort Harker.

Frontiers Defense By Buffalo Speakers

April 1867 There was a serious need for protection and defensive measures when the Indians were consistently raiding and stealing cattle in April 1867. The commanding officer of the Seventh Cavalry Regiment dispatched Company D, E, and I near Santa Fe routes. They were given tasks to guard and protect railroad, crews and provide escort duties.

Skirmish with Cheyenne warriors

September 15, 1868, Sergeant Davis of the 10th Cavalry was in command of the detachment of nine men. They were attacked by some 60 Cheyenne warriors while on patrol duty. They were successful in forcing the Indians to flee. One trooper was wounded and two railroad laborers killed.

Some Activities of the Tenth Cavalry

1868   The Tenth Cavalry Regiment had units dispersed throughout Kansas and the Indian territory in 1868. The companies were patrolling the Union Pacific Railroad areas. Some Buffalo Soldiers participated in the campaigns against Black Kettle and his Indian warriors. The regiment lost a considerable number of horses when they were in a blizzard. On September 15, 1868 Company I was involved in a skirmish with Indian warriors. They lost ten horses and killed seven Indian warriors.

Use of Sterotypes by Buffalo Bill Cody

1868   A book on Buffalo Bill Cody stated that when he arrived at Fort Hays and reported to the Commanding Officer, Captain Graham, he was told that some Buffalo Soldiers would be going on a pursuit mission to locate some Indian warriors. Buffalo Bill responded "Darkies have never been in an Indian fight." While on a mission with Captain Graham's Company, near the Saline River Buffalo Bill Cody said the following occurred. One of the troopers mistakenly fired his weapon and alerted the Indian warriors causing them to flee. Cody said "Captain Graham was bitter and told the nigger who fired the gun, that as punishment for his carelessness, he had to walk all the way back to Fort Hayes," These statements were the true character as far as racial views are concerned for one of America's popular Western hero. (to some admirers).

Tenth Cavalry Soldiers to the Rescue, Battle of Beechers Island

September 1868   When Colonel George A. Forsyth and some civilian scouts were attacked by 700 Indian warriors on the island in the Republican River, the Tenth Cavalry came to the rescue. Buffalo Soldiers of Companies, H, and I, assisted Forsyth and the scouts.

A Brief Skirmish With Indians

1868   Captain Henry Alvord and one company of the Tenth Cavalry and a detachment of the Sixth Infantry Regiment departed Fort Arbuckle to provide protection at the Fort Cobb Indian Agency.

Alvord's party confronted some Kiowa and Comanche Indian warriors at Cottonwood Grove, Texas. After a brief stand off, they continued their march toward Fort Cobb.

Buffalo Soldiers Build Structures at Camp Wichita, Indian Territory

April 1869 The Tenth Cavalry Regiment was transferred to Camp Wichita in April 1869. They were given the mission to construct buildings at the camp located on "a flat northeast side of a site selected for a permanent Post, on Medicine Bluff Creek near its junction with Cache Creek". The soldiers used an old sawmill and obtained logs from the Wichita Mountains. Temporary buildings were constructed consisting of materials from rock quarries. The floors were of mud. They called the buildings "jackal". Later construction was started on a permanent post. The officers supervised the construction and the noncommissioned officers (NCO) were the foremen. These Real Buffalo Soldiers, many who could not read or write were able in 1869 to "square logs with broad axes, quarry rock, burn lime, dress lumber and lay stone". These talented troops built barracks for ten troops of cavalry, officers quarters, stables and storehouses. Some of these original buildings were still standing at Fort Sill Oklahoma in the 1960's and probably are today in 1994.

Wild Bill Hickok and Tenth Cavalry

1869 The movies and magazines of the West today do not reveal the fact that in 1869, the celebrated Wild Bill Hickok was present at Fort Lyons and was present with the 10th Cavalry for a brief period during General Sheridan's winter campaign.

White Thieves Steal Mules

1869 When some white outlaws stole 139 government mules in Indian Territory, the Buffalo Soldiers chased them for two days. They were able to arrest 7 outlaws and recovered 127 mules.

A Skirmish with Satanta warriors

January 1870 John Harshfield, a meat contractor for the government, was bringing a herd of cattle for the Cheyenne and Apache Indians at the Indian Agency near Camp Supply. Harshfield and 14 men were involved in a skirmish with Indian warriors from Satanta's band. The attack was finally stopped due to the actions by an Indian leader, Kicking Bird. He was able to influence Satanta's Warrior to stop the attack. A Major Meredith Kidd, Commander at Camp Supply and four companies of the 10th Cavalry rescued the Harshfield party and chased the Indians until they crossed the Canadian river.

Rescue of White Children

February 1870 On February 25, 1870, Company D, 10th Cavalry marched from Fort Arbuckle to Cottonwood Grove, Texas to assist in the rescue of some white children held captive by some Indian warriors.

Buffalo Soldiers Escort Generals

1870 When some general officers and high ranking officers would visit in the Indian Territory, Buffalo Soldiers of the 10th Cavalry would provide escort details. They would provide escorts during visits to Forts McKavett, Concho, Griffin and Fort Richardson in Texas.

A Skirmish with Indian warriors

1871 On June 11, 1871, a party of Comanche Indian warriors stampeded some horses at Camp Supply. The Indians were pursued by 10th Cavalry's companies A, F, H, I, K, and three companies of the Third Infantry Regiment Six Indians were killed and three soldiers wounded.

Buffalo Soldiers Arrest Whiskey Ranches

1872 Twenty Buffalo Soldiers of Company D were dispatched to arrest some whisky ranchers who operated whiskey ranches. The soldiers arrested 15 men and destroyed many gallons of whiskey.

32 The Real Buffalo Soldiers

1872 When General Sherman, Army Commanding General was visiting the Department of Texas in 1872, he observed a mule supply train moving toward Fort Griffin. Upon his arrival at Fort Richardson, he learned that the mule supply train had been attacked by Indian warriors. Sherman continued on to Fort Sill, Indian Territory. A military reservation a half a mile from Fort Sill had an Indian Agency that was visited by some seven or eight thousand Indians, men, women and children. They were the Kiowas and Comanches. A Kiowa Chief Satanta, was bragging about his raiding the mule supply wagon train. The Indian agent had Satanta, Satank, and Big Tree arrested and brought to Fort Sill, In preparation for the visit of the Indian leaders with General Sherman, the fort commander, had his troops prepared for any problems that could occur. There were troops on horses hidden in the stables, some soldiers were posted in the commanders' house, hidden behind some window shutters with rifles prepared for action while the Indian chiefs were talking to General Sherman on the commander's porch. Big Tree rode by the house on his way to a traders store. The adjutant Lieutenant Woodard and some soldiers were preparing to arrest him when he wrapped his blanket around his head and leaped through the window. He vaulted a high fence and ran through the gardens. Finally he was captured. Just as General Sherman gave a signal for the concealed soldiers to arrest Satanta and Satank, Lone Wolf, a Kiowa chief rode up. He dismounted and opened his blanket displaying two loaded carbines. He gave one to another Indian Warrior. The Tenth Cavalry's commanding officer Colonel Grierson leaped upon the two Indians, grabbed their weapons and was able to throw the men to the ground. There were some Indians who became excited as the Cavalry troops were moving about in the compound. However the soldiers were able to maintain order. The Fourth Cavalry regiment was given the task to turn the Indians, Satanta, Satank, and Big Tree over to Texas civil authorities. Satank began a weird chant and Big Tree and Satanta grabbed Satank and made him get into the wagon. Satank rode in a wagon with a fourth cavalryman. Satank continued his chanting and a few miles away from the post, Satank drew a knife and stabbed the soldier. He then threatened some guards with a loaded rifle. Satank was shot through the wrist and was killed by a second bullet.

Attack Upon Wichita Agency

1874 On August 23, 1874, some Indians who did not like the military post at the Wichita Agency and attacked the gunman. They were a band of Kiowas and Naconees. They had a skirmish with Companies C, E, H, and I of the Tenth Cavalry.

Battle At Elk Creek

1874 The Buffalo Soldiers of Companies D and M along with one Company of the 11th Infantry Regiment under command of Major Schofield were present at Elk Creek on October 25, 1874. They were successful in capturing 300 ponies. The prisoners were taken to Fort Sill and later sent to Fort Marion, Florida.

Pursuit of Black Horse's warriors

1875 While awaiting transportation from the Cheyenne Agency to Fort Marion, Black Horse on April 6, 1875, knocked the guard down and ran toward his tribe's camp. After Black Horse was killed, some members of his tribe fled toward the hills. They became involved in a skirmish with Buffalo Soldiers of Companies D and M of the Tenth Cavalry. The 10th Cavalry had one trooper killed and eight wounded.

Relocation of Tenth Cavalry Regiment

1875 The 10th Cavalry Headquarters was transferred to Fort Concho on April 17, 1875. Other troops were stationed at various forts. Companies A, D, F, G, I and L at Fort Concho, Companies B and E at Fort Griffin, Companies C and K at Fort McKavett, Company H at Fort Davis and Company M at Fort Stockton, Texas. The Tenth Cavalry regiment's "Real Buffalo Soldiers" were also called "the wild buffalos".

A Surprise at Saragossa

1876 A detachment of Buffalo Soldiers under the leadership of Lieutenant Evans and with the assistance of two Seminole scouts surprised a camp of Lipan and Kickapoo Indians near Saragossa, Mexico on July 30, 1876. This was an accomplished feat for the soldiers because they had marched 110 miles in 25 hours. Ten Indians were killed and 93 captured. The 10th Cavalry lost one horse and 23 lodges were destroyed.

Indian Village Destroyed

On August 12, 1876, Captain Lebo and soldiers of Companies B, E, and K destroyed a village of hostile Indian warriors.

Capture of Mexican horse Thieves

November 1876   A detachment of Buffalo Soldiers from Company G, Fort Griffin, Texas marched a distance of 770 miles. They pursued some Mexican horse thieves. Corporal John Robinson and four enlisted men captured 10 Mexicans and 15 horses.

Relief of Captain Nolan

1877   First Lieutenant R. G. Smithers, adjutant of the Tenth Cavalry along with 16 Buffalo Soldiers marched to the rescue of Captain Nolan's command on August 3. They traveled a distance of 140 miles in 41 hours. Nolan's men had been fighting the Indians under extreme weather conditions. There were men dying from lack of water. When the detachment returned to Fort Concho on August 14, 1877, they had traveled a total distance of 1500 miles.

Pursuit of Commanche warriors

1877   On April 9, 1877, a detachment of 42 Buffalo Soldiers under the command of Captain Lee and Lieutenant Jones left Fort Griffin, Texas to pursue some Indian warriors. They surprised some Comanche Indians at their village, Lake Quemado, Texas. Four Indian warriors were killed and six squaws captured. A first Sergeant Charles Baker was killed in action. The detachment had marched a total distance of 750 miles.

Return From Scouting Mission

1877   Twenty four enlisted men and one officer, 10th Cavalry Regiment returned to Fort Richardson, Texas after traveling a total distance of 1,360 miles on the staked plains.

Skirmish with Mescalero Indians

1880 Company K Tenth Cavalry had marched through the Guadalupe Mountains by way of Guadalupe Creek to the Rio Panaso in the Sacramento Mountains to the Mescalero Indian Agency to disarm them. On April 9, 1880, the Buffalo Soldiers entered the camp of a small group of Mescalero Indians at Shakehand Springs, New Mexico. One Indian was killed, four women captured and they released one chief from captivity and a small Mexican boy named Cayetarra Segura, 11 years old. The soldiers marched a distance of 417 miles. They also captured 21 horses and mules and destroyed the camp.

A Pursuit Mission

1880 On August 2, 1880, Company A Tenth Cavalry began a pursuit march for some Indian warriors. They marched from the area of old Fort Quitman to Vanhorn Wells on to Devils Race Course, to Rattle Snake Springs, into the passes of Sierra Diablo. On August 11, after arriving at Ash Springs the soldiers began to follow the trail of Victorio's Apache warriors. The Indians made it to the Rio Grande and then cross the river into Mexico. The Buffalo Soldiers marched a distance of 748 miles.

A Skirmish at Eagle Spring, Texas

1880 Company G, Tenth Cavalry departed for Sulphur Water Hole, Texas, August 1880 and marched to Van Horn, Texas and to Rattlesnake Springs. On August 6, 1880, the Buffalo Soldiers confronted some Indian warriors near Rattlesnake Springs. The company arrived at Sulphur Water Hill, Texas on August 7, 1880. Company H was engaged in "furnishing pickets and scouts from Eagle Springs, Texas." On August 3, Corporal A. Weaker and Private Brent of Co. A and a small detail from some other companies were on picket duty at Alamo Springs and sighted some of Victoria's warriors as they crossed the Rio Grande. There was a skirmish that involved a running fight for 15 miles. On August 3, 1880, the company left Eagle Springs to follow the Indian warriors. They marched to Devil's Race Course and on to Rattlesnake Springs. Company H found Companies B, C, and G on August 6, 1880, and engaged in a fight with Victoria's warriors. The Indians finally retreated toward the mountains. The Buffalo Soldiers had marched a distance of 1,250 miles.

The Capture of Victoria

1881 The Tenth U.S. Cavalry persistent campaigns and expeditions to arrest Victoria and his Indian warriors contributed to the final flight across the border by Victoria and his warriors. Because in October 1881, a combined force of American Mexican military decided to wage an all out campaign against Victoria. They led a large force into Mexico. The Mexican commander was Colonel Juaquin Terraza. Ten companies of the 10th Cavalry under the command of Colonel Grierson were located along the Rio Grande, to prevent Victoria from entering Texas. Victoria and his warriors were in the area of Tres Castillos Mountains in Mexico. On October 14, 1881, the Mexican soldiers attacked from all sides as they approached Victoria's possession. Victoria was killed along with some of his Indian warriors.

"A Change of Name"

1881 On January 1, 1881, the term "troop" was used when troops were reorganized from companies. The Cavalry soldiers were officially called troopers.

ARRIVAL AT FORT DAVIS, TEXAS

1882 The Tenth U.S. Cavalry Regiment Headquarters transferred to Fort Davis on July 1881, where it remained until March 1885.

The Pursuit of Geronimo and His warriors

1885 The Tenth US Cavalry Regiment relocated to the Department of Arizona in 1885. The Regiment was given a mission to follow the trails of Indian warriors, "Geronimo, Apache Kid, Mangus, Cochise Alchise, Aklenni, and Natsin Eskiltie" who were committing serious raids and murders. The troops of the 10th Cavalry were involved in numerous confrontations and skirmishes with these Indian warriors, especially Geronimo's band of warriors.

Disposition of Tenth Cavalry Troops

1885 The 10th Cavalry Regiment was located briefly together as a composite unit at Camp Rice. Later the troops were assigned stations as follows: "Headquarter and Troop B, Whipple Barracks, Troop A, Fort Apache; C, F, and G, Fort Thomas; D, E, H, K, and L, Fort Grant; I and M, Fort Verde."

Chasing The Indian warriors

1885 Major Van Vliet and Troops D, E, H, K departed Fort Thomas to chase some of Geronimo's Indian warriors. The Buffalo Soldiers marched to the Mongolian Mountains near Fort Bayard, New Mexico. They were unsuccessful in finding the Indian warriors.

A Heroic Officer

1886 The Buffalo Soldiers of Troop K and their commander, Captain Lebo and Lieutenant Powhattan H. Clark, had just completed a 200 mile march from Calabooses into Mexico. On May 3, 1886, the soldiers engaged in a fight in the Pinto Mountains with Geronimo's Indian warriors. The Apaches were able to maintain a strategic strong hold on the gorges and cliffs of the mountains until their retreat. Lieutenant Powhattan Clark demonstrated a heroic act when he went to the aid of a wounded trooper who was still exposed to enemy fire. Clark was able to rescue Corporal Winfield Scott by carrying him to safety. Lieutenant Clark was awarded the Congressional Medal of Honor for his heroic performance.

A Tenth Cavalry Victory

October 1886 The Buffalo Troopers or the "Wild Buffalos" were successful in defeating Chief, Mangus and his Indian warriors in the White Mountains east of Fort Apache.

The Finale of the Indian Campaigns

1886 The "Real Buffalo Soldiers" of the Tenth U.S. Cavalry Regiment had performed their specific duties and missions in a most outstanding manner from 1866 to 1886. They were confronted with problems of inferior equipment, second hand items at times, not the best horses and mules, had to construct buildings, prepared camp sites and engaged in many skirmishes and battles with the Indian warriors under extreme conditions. The 10th Cavalry Troopers marched thousands of miles through rough terrain and under severe weather conditions. They also experienced the insults, acts of racism and the exposure to segregated conditions sometimes precipitated by the white settlers they were assisting and defending from Indian warriors. Upon their arrival in Santa Fe, New Mexico, a general order had been issued, dated October 7, 1886. The general order announced the ending of the Indian Campaign. The outstanding commanding officer of the 10th Cavalry was promoted to Brigadier General, Benjamin Henry Grierson. He was the first Commanding officer of the Tenth U.S. Cavalry Regiment. He was reassigned upon promotion.

On To Montana

1890 The new commanding officer of the Tenth U. S. Cavalry Regiment, Colonel J. K. Mizner had requested a new location for the regiment. The regimental Headquarters was moved to Fort Custer, Montana. Even in 1890, the regiment would receive second hand horses. The 10th Cavalry relieved the First Cavalry Regiment in Montana and obtained their horses. The troops were assigned to the following stations: "Band, Troops A, B, E, G, and K, Fort Custer, Montana; Troops C and F, Fort Assinniboine, Montana, Troop D, Fort Keogh Montana; Troop H, Fort Buford, North Dakota.

Truant Officer Duty

1891 In 1891, some Buffalo Soldiers left Fort Apache to march to an Indian reservation, Orabi Village. The troops were enforcing compulsory school attendance for Indian children.

Demise of Lieutenants Clark and Finley

1893-1894 There was sorrow in the Tenth Cavalry Regiment on July 21, 1893, when Lieutenant Clark drowned in the Little Big Horn River and in February, 1894 when Lieutenant Finley died after amputation of his leg. His leg was crushed when a horse fell on him.

Protection for the Railroad

1894   The Buffalo Soldiers of Troops B, E, G, and K were ordered to maintain law and order in April, 1894 when some strikers of Coxey's Army had stopped a Northern Pacific Train. It was necessary for some troops to be detailed as railroad guards.

Lieutenant John "Black Jack" Pershing with The Tenth Cavalry Regiment

1896   The Tenth Cavalry Regiment's troops were given a mission to return some Cree Indians to their Canadian reservation. The Indians had left the reservation and had been involved in minor offenses since 1886. Lieutenant John J. Pershing was the commander of Troop D and his troops spent the summer on this mission. The soldiers had marched 600 miles.

THE ARREST OF SOME CHEYENNE INDIANS.

1897   In 1897, Buffalo Soldiers of Troops A, E, and K were ordered to arrest some outlaw Cheyenne Indians at the Tongue River Agency. Some of the Indians arrested were "Yellow Hair, Whirlwind, Should Blade, and Sam Crow".

THE TENTH CAVALRY IN CUBA, 1898.

1898   I can recall that as a young boy, at the age of ten years, I was visiting my father who was a physical education director at the Washington, D.C. 12th Street YMCA and I was introduced to a man who was a veteran of the Spanish American War. I observed the senior citizen playing checkers and smoking his pipe. I did not know what the Spanish American War was all about and also the Tenth Cavalry Regiment. I pose the question, how many youngsters of all ages and races know today about those famous Buffalo Soldiers of the Spanish American War? The Spanish American War played a great part in the military history of the African American's combat efficiency and capabilities. The Tenth Cavalry Regiment beyond any doubt along with the other three Buffalo regiments proved to white and black Americans and to the world that people of color, black, brown and yellow complexions called Negroes, black and African American could and did fight in combat and performed heroically and outstanding under extreme combat conditions. Unfortunately, this was not considered in June 1944 for D-Day in Europe. Yes, in the Spanish American War, those proud "Tenth Cavalry Buffalo Soldiers accomplished these exploits. The 10th Cavalry participated in the capturing of Kettle Hill. Sergeant George Berry, 9th Cavalry's color bearer took the colors from the Third Cavalry's wounded color bearer. Certificates of merits and distinguished awards were given to troopers Corporal Walker and Private Smith. Also receiving commendations were: Troop A, Corporal John Anderson, Private R. A. Parker, Troop C, Sergeant Adam Houston, Troop E, First Sergeant Peter McCan, Sergeants Benjamin Fasit, O.G. Gaither, William Payne, Corporal Thomas H. Herbert, Troop I, Private Elsie Jones.

Regiment Return to U.S.

1899 The Tenth U.S. Cavalry Regiment's troops were assigned to the following locations in Texas: Headquarters A,G,H and L, at Fort Sam Houston, Texas; C,D and M At Fort Clark; B at Fort Ringgold; E at Fort McIntosh, F at Eagle Pass; I at Ft. Clark and K at Fort Brown.

In February 1895, the 10th Cavalry Regiment returned to Cuba. The regiment was stationed at Manzanillo, Troops A, C and H, Lat Bayamo, M at Jighuani, C at Campechuela, B at Gibra, D, E and I at Holguin, Fort Banes, and K at Porto Padre. During their stay, the 10th Cavalry confronted some bandits and outlaws. "Lieutenant Walter C. Short captured eleven outlaws and the Cuban bandit Troneon was arrested.

Tenth Cavalry Participation in Philippine Insurrection

1901 The Second Squadron, 10th Cavalry returned to Texas from Cuba in January 1900. On April 1, 1901, the squadron was ordered to the Philippines. The squadron was stationed at Samar, Bibotan, Gandara, Calbayog and Panay.

Tenth Cavalry and The Utes

1906 The Tenth Cavalry Regiment was ordered to assist the Sixth Cavalry Regiment in quelling an uprising of the Ute Indians on a Wyoming, Montana Indian Reservation. The Utes were demonstrating and terrorizing outside of the reservation. Eight troops of the Tenth U.S. Cavalry Regiment were successful in having the Utes to return to their reservation.

Tenth Cavalary's Trip Abroad

1909   In 1909, the Buffalo Soldiers had traveled on a voyage and visited Singapore, Columbo, Aden, Suez, Port Said, Alexandria, Valletta and Gibraltar. The Buffalo Troop returned to New York on July 25, 1909 after traveling 10,729 miles. The regiment then was stationed at Fort Ethan Allen, Vermont.

Buffalo Soldiers Wins First Place In Riding Competition

1913   A corporal of Troop C, Tenth U.S. Cavalry Regiment won first place in the soldier's horse jumping class during a horse show in Washington, D.C.

Chief Musician Receives Scholarship

1914   Chief Musician Alfred J. Thomas, Tenth U.S. Cavalry Regiment was the recipient of a scholarship to attend the Juillard School of Music on a two year full scholarship.

Tenth Cavalry in Mexico

1913-
1914   Upon receipt of orders for the Tenth U.S. Cavalry to participate in the eventual Punitive Expedition to Mexico, the regimental commander sent an advance guard, Troop L, on November 27, 1913 to Fort Apache. They arrived on December 6, 1913. The remaining regiment arrived at Fort Huachuca, on December 19, 1913. Troops were sent to patrol along the border where unrest was evident. Troop E was at Naco, G,H and M, Nogales, K, Forrest and Troop A had a detachment at Yuma. The Tenth Cavalry was present during the seize of Naco or confrontations between the "Carrancistas and Constitutionalistas". Eight men of the 10th Cavalry were wounded. Later the Secretary of War, Lindley M. Garrison sent a letter to the regiment. The letter stated that "*By direction of the President, I take great pleasure in expressing to the officer and enlisted men of the Tenth Cavalry his appreciation of their splendid conduct and efficient service*

in the enforcement of the United States Neutrality laws at Naco, Arizona during November, December and January last."

1916 On March 9, 1916, Colonel W.C. Brown, commanding officer, Tenth U.S. Cavalry issued orders for his regiment to prepare for action in Mexico. Pancho Villa had made a raid in Columbus, New Mexico, and several soldiers and civilians were killed. Orders came immediately from cavalry brigade headquarters at Douglas, Arizona.

The Tenth Cavalry Regiment crossed the border into Mexico on March 15, 1916. During their stay in Mexico in the pursuit of Pancho Villa and his Villistas, the regiment traveled through the following towns Colonia Dublan, El Rio Casa Grandes, La Joya, El Rucio, San Miguel, El Toro, La Osa, Sapien, Cusihirachic, Auguas, Calientes, Parral, Santa Cruz de Villagas, Ahumada, and Valle de Zaragoza.

During the Punitive Expedition in Mexico, Colonel Charles Young (then Major) led his Buffalo troopers to the aid of Major Tompkins and his cavalry unit. A Captain Boyd and Lieutenant Adair and some Buffalo Soldiers were killed by the Mexicans at Carrizal. The remaining members of Boyd's Troop C were forced to retreat. The 10th Cavalry Regiment returned to the United States on February 5, 1917. It is believed that a Corporal John A. Jeter Jr., demonstrated his leadership by assuming command of his unit when his commanding officer was killed on June 21, 1916.

World War I

1918 During World War I, the Tenth Cavalry was stationed along the Mexican border on patrol duty. There were 64 noncommissioned officers from the Tenth Cavalry who left the regiment and were later commissioned as officers and assigned to other units.

Tenth Cavalry and the Yaqui Indians

1918 On January 9, 1918, 30 Yaqui Indians opened fire on soldiers of Troop E, 10th Cavalry regiment at Atasco Canyon, west of Nogales, Arizona. The Buffalo Soldiers responded to the Indians hostile actions.

Tenth Cavalry and Mexican Soldiers

1918 In August 1918, a battle occurred on the border line between some American Customs men and Mexican soldiers. An officer of the Tenth Cavalry, Lieutenant Colonel (LTC) Herman dispatched some Buffalo Soldier to Nogales and three companies of Infantry were also sent to Nogales. The Mexican soldiers were firing into the town and when Captain Hungerford of Troop C was ordered "to cross the line and clear the hill of Mexicans", he was killed in the attack. Other officers wounded were LTC Herman, he was wounded in the leg and Captain Caron wounded in the right arm." A truce was finally reached after 129 Mexicans were killed. It is believed that two German army officers were responsible for the fighting or skirmish that occurred.

Disposition of the Tenth Cavalry Regiment

1929 The Tenth Cavalry Regiment was patrolling along the Mexican border from Douglas through Nogales and as far west as San Miguel, Arizona. When the Mexican Revolution was over, the regiment returned to Fort Huachuca, Arizona.

1931 The 10th Cavalry Regiment was located in 1931 at: Regimental Headquarters, Headquarters Troop and First Squadron, Fort Leavenworth, Kansas, Second Squadron to the United States Military Academy, West Point, New York, and the Machine Gun Troop to Fort Myer, Virginia. Troop F at West Point exchanged places with the Machine Gun Troop."

1941 On April 1, 1941, the Tenth U.S. Cavalry Regiment was at full strength, a total of 1,326 men, stationed at Camp Funston, Fort Riley, Kansas.

Tenth U.S. Cavalry At Camp Lockett, California

1943 In August 1994, some areas of California were experiencing some serious and damaging forest fires. Civilian and military personnel were fighting those fires. The year was 1943 and those Buffalo Soldiers of the Tenth U.S. Cavalry Regiment and a recently activated black cavalry regiment, the 28th regiment departed Camp Lockett, California to assist in fighting some forest fires. A supervisor of the National Parks wrote to the commander of the Cavalry Brigade:

"There is no telling how large these fires might have been had it not been for the rapid mobilization of your men from Camp Lockett".

Buffalo Soldier Received Commendation

April 1
1943 A general order number 3, published by Headquarters, Tenth U.S. Cavalry Regiment, Camp Lockett, California, dated April 1, 1943 commended Private Hosea B. Smith, Troop B, Tenth Cavalry regiment. The regimental commander praised Smith for his attempt to halt a runaway team of horses. *"On March 29, 1943 Private Smith was driving an untamed four-line team of horses back from the drill field when the team broke out of control. Private Smith was thrown from the wagon but retained the reins in his grasp and was dragged for about three hundred yards before the team escaped."* The commendation also stated that *"This courageous effort to halt a runaway team shows Private Smith to be a man possessed of judgement, courage and a most commendable sense of duty, all exercised without regard to his own personal safety. His actions on this occasion is a fine example for all of us.*

In the late 1950's, I would daily past the large riding halls and horse stables on my way to and from work as the engineer officer, United States Disciplinary Barracks, Fort Leavenworth, Kansas. I was told that many years ago a very distinct and proud regiment of men called the Tenth U.S. Cavalrymen were stationed here and used those facilities still standing and at times were maintained and repaired by the post engineer facilities.

I had no crystal ball or belief in the future that I would take this pen in hand and write some true facts about those "Real Buffalo Soldiers of the Tenth Cavalry regiment. The halls are silent, and the stables are absent of the presence of those Buffalo Soldiers' horses, but their memories have been resurrected with a memorial stamp and people selling memorabilia and other artifacts to represent the Buffalo Soldiers. However, I must state it very frankly that many citizens of America, especially the ancestors of the white majority, have profited in their safety, survival and future security from the efforts of those men who were called Niggers, Brunettes, darkies and Buffalo Soldiers, but really they were the "Black Defenders of America from 1866-1943".

"THEY WERE THERE IN BATTLES"

It is unfortunate that in July 1994, television weekly serials still must depict black characters in a somber, pacifist role to justify the intent of the script. A very interesting episode of a popular television show about a daring female doctor in the west featured the Buffalo Soldiers. The black sergeant was portrayed as very concerned about the Indian's plight and found it necessary to leave the regiment. I will agree that there were many soldiers, white and black in cavalry regiments who had concern for the humanitarian part of the Native American and could have deserted or resigned from their regiments. As I have stated in this manuscript that there were many reasons why a second class citizen in America - the black man enlisted in the U.S. Army. The picture does not relate the true entire story of the black presence in the west as cavalrymen. Because I am sure that when those Real Buffalo Soldiers were present in numerous engagements of the regiment, that they did have second thoughts about the enemy they were facing. They probably were confused as to how an Indian's genetic brother could fight for the white man as Indian scout. Yes, they were there. African Americans as Buffalo Soldiers of the Tenth Cavalry Regiment were present in these engagements of their regiment:

"*Saline River, Kansas, August 2, 1867, Co. F, Cheyennes.*

Near Saline River, Kansas, August 21, 1867, Co. F, Cheyennes

Fort Hays, Kansas, September 15, 1867, Co. G, Cheyennes

Big Sandy Creek, Kansas September 15, 1867, Co. I Cheyennes

Beaver Creek, Kansas, October 18, 1867, Companies H and I, Comanches

Camp Supply, Indian Territory, June 11, 1871. Companies A,F,H,I and K, Comanches

Double Mountain, Indian Territory, February 5, 1874, Companies D and G, Comanches

Wichita, Indian Territory, August 22, 23, 1874, Headquarters, Companies, C,E,H and L, Kiowas and Comanches

Buffalo Springs Indian Territory, April 6, 1875, Co. M, Cheyennes

Sulphur Springs, Texas, July 6, 1875, Co. A, Comanches

Saragossa, Mexico, July 30, 1876, Co. B, Apaches

Pinto Mountain, Mexico, August 12, 1876, Companies B and K, Apaches

Peco Springs, Texas, September 13, 1876, Co. G, Comanches

Near Fort Griffin, Texas, May 4, 1877, Co. G, Comanches

In Mexico, September 29, 1877, Co. C, Comanches

Sierra Cannel, Mexico, November 29, 1877, Co. B, Alsatti's Apaches

Salt Lakes, Texas, July 29, 1879, Co. H, Comanches

Pecos River, Texas, April 2, 1880, Co. L, Comanches

Shakehand Springs, Texas, April 30, 1880, Co. K, Comanches

Tinaja De Las Palmas, July 30, 1880, Headquarters and Band, Apaches

Eagle Springs, Texas, Jul 30, 1880, Companies C and G, Apaches

Alamo Springs, Texas, August 3, 1880, Company H, Apaches

Near Camp Safford, Texas, August 7, 1880, Companies F and L, Apaches,

Rattlesnake Springs, Texas, August 7, 1880, Companies, B,C,G and H, Apaches

Ojo Caliente, Texas, October 28, 1880, Co. B, Comanches

Pinto Mountains, Mexico, May 3, 1885, Troop G, Geronimo's Apaches

White Mountains, Arizona, September 18, 1886, Troop H, capture of Mangus. Las Guasimas, Cuba, June 24, 1898, Troops A,B,E and I Santiago, Cuba. July 1,2,3, 1898, Troops, A,B,C,D,E,F,G, and I Carrizal, Mexico. June 21, 1916, Troops C and K Nogales, Arizona. August 27, 1918, Troops A,C, and F".

They Rest Among The Known

We should not forget, history cannot forget, therefore we must remember in this day and time that there were some brave, courageous, and valiant Buffalo soldiers and their officers who lived up to the Regimental Battle Cry: "Ready and Forward". As these men moved forward, they were stopped in battle. However, their names inscribed upon a Roll of Honor and there silent responses to the call of taps will always be remembered. Those Real Buffalo Soldiers who were killed in action are:

"Sergeant (Sgt.) William Christy, Company (Co.) F, Saline River, Kansas, August 2, 1867

Private (Pvt.) Thomas Smith, Co. F, Near Fort Hays, Kansas, August 21, 1867

Wagoner Larkin Foster, Co. B, Foster Springs, Texas, September 19, 1871

Pvt. Clark Young, Co. M, Cheyenne Agency, Indian Territory, April 12, 1875

First Sergeant Charles Butler, Co. G, Lake Quemado, Texas, May 4, 1877

Pvt. Martin Davis, Co. C, Eagle Springs, Texas, July 30, 1880

Pvt. William Taylor, Co. F, Camp Safford, Texas, August 4, 1880

Pvt. Wesley Hardy, Co. H, Rattlesnake Springs, Texas, August 6, 1880

Pvt. George Locks, Co. C, Eagle Springs, Texas, August 30, 1880

Pvt. Carter Burns, Co. B., Ojo Caliente, Texas, October 28, 1880

Pvt. George Mills, Co. B, Ojo Caliente, Texas, October 28, 1880

Corporal (Cpl.) William Backers, Co. K, Ojo Caliente, Texas, October 28, 1880

Pvt. J. K. Griffin, Co. K, Ojo Caliente, Texas, October 28, 1880

Pvt. James Stanley, Co. K, Ojo Caliente, Texas, October 28, 1880

Pvt J. Follis, Troop K, Pinto Mountains, Texas, May 3, 1886
Sgt. Robert Evans, Troop C, Gaileyville Canyon, June 3, 1886

Cpl. William L. White, Troop E, Guasimas, Cuba, June 24, 1898

Pvt. John H. Smoot, Troop A, San Juan, Cuba, July 1, 1898

Cpl. William F. Johnson, Troop B San Juan, Cuba, July 1, 1898

Pvt. John H. Dodson, Troop C, San Juan, Cuba, July 1, 1898

Pvt George Stovall, Troop D, San Juan, Cuba, July 1, 1898

Pvt William H. Slaughter, Troop G. San Juan, Cuba, July 1, 1898

First Sergeant William Winrow, Troop C, Carrizal, Mexico, June 21, 1916

Sgt. Will Hines, Troop C, Carrizal, Mexico, June 21, 1916

Pvt. Thomas Moses, Troop C, Carrizal, Mexico, June 21, 1916

Horseshoer Lee Talbot, Troop C, Carrizal, Mexico, June 21, 1916

Pvt. DeWitt Rucker, Troop K, Carrizal, Mexico, June 21, 1916

Pvt James E. Day, Troop K, Carrizal, Mexico, June 21, 1916

CHAPTER 3

TWENTY-FOURTH INFANTRY REGIMENT

Twenty Fourth Infantry Regiment

The 1866 Act involving the reorganization of the regular United States Army included a provision for four black Infantry regiments. However in 1859, they were reduced to two. The Twenty-fourth Infantry Regiment were created by the consolidation of the Thirty-eighth Infantry Regiment with the Forty-first Infantry Regiment. The Twenty-fourth Infantry Regiment was inactivated in Korea on October 1, 1956.

Where Were They?

The "Real Buffalo Soldiers" of the Twenty-fourth Infantry Regiment were present in the "Winning of the West", Spanish American War, and the Philippine Insurrection. Unfortunately, the following historical facts are still missing from many secondary and college text books.

They Were There
Reinforcement Mission

1867   **Reinforcement Mission.** A detachment consisting of 25 men of the Twenty-fourth Infantry Regiment were used as reinforcements at one time for a cavalry unit in distress. The post commander of Fort Hayes had ordered the men to mount up (use horses) and bring along their howitzer weapons to fight the attacking Indian warriors.

Twenty-fourth with Shafter's Expedition

June 1871   Lieutenant Colonel (LTC) William R. Shafter led an expedition of soldiers from Barilla Springs, Fort Davis, and Fort Stockton, Texas. A detachment of the Twenty-fourth Infantry Regiment was with the expedition. They traveled to the Pecos, White Sands region.

Patrolling Duties

1871   The Twenty-fourth Infantry Regiment had duties of patrolling areas during the Indian campaigns. The brave Buffalo Soldiers patrolled the Carrigo Mountains, Alamo Springs, Eagle Springs, and Van Horn Wells Mail Station. Sometimes the soldiers confronted Indian warriors.

A Searching Mission

1873 The Twenty-fourth Infantry soldiers were present in field operations during the Indian Wars. On May 16, 1873, Colonel MacKenzie departed from Fort Clark, Texas to search for some Indian warriors. His search force consisted of six companies of his regiment and a detachment of the Twenty-fourth Infantry Regiment.

Pursuing Indian warriors

1875 Colonel Shafter commanded an expedition that left Fort Concho on July 14, 1875 to pursue some Indian warriors. His command consisted of six companies of the Tenth Cavalry and two companies of the Twenty-fourth Infantry. There were also some Seminole and Tonkawa Scouts.

Pursuit Mission

July 1876 Lieutenant Bullis and soldiers of the Twenty-fourth Infantry marched 110 miles in 23 hours and were successful in surprising a camp of 23 Indian lodges. The Indians were from the Lipans and Kickapoo tribes, who were camped near Saragossa, Mexico. There were some Indian casualties and the soldiers captured many horses.

Searching for Victoria

1880 During the Victoria War, Colonel Grierson, commanding officer, Tenth U.S. Cavalry Regiment was searching for Victoria and his Indian warriors near Rattlesnake Springs around July 30, 1880. Company H of the Twenty-fourth Infantry Regiment was part of the search force.

Supply Train Escorts

1880 An army supply train was being escorted by some members of the Twenty-fourth Infantry. A band of Apache warriors approached the train and received an element of surprise, because hidden inside of the wagon train were the soldiers. They immediately commenced firing and were able to force the Indians to retreat. This is an example of the fact that these

Infantry men were present during the Indian Wars and sometimes became involved in skirmishes just as the cavalry men did on various occasions. They were Buffalo Soldiers also.

Presence of a Chaplain of Color

1885 While the regiment was assigned along the borders of Arizona, there was present a black Chaplain, Chaplain Allen Allensworth. (See biographical sketch)

Participation in Equipment Experiments

1893 Members of the regiment were involved in testing an experimented blanket roll for the army.

Citizens Say Farewell

1898 The citizens of Missoula, Montana showed their appreciation for the soldiers of the Twenty-fourth Infantry Regiment when they departed for Cuba, during the Spanish American War.

1898 The Twenty-fourth Infantry Regiment provided some support for the Tenth Cavalry Regiment during the battle at San Juan Hill, Cuba. Lieutenant Lyon of the Twenty-fourth Infantry and his Buffalo Soldiers provided a hotchkiss gun in support of the advancing Tenth Cavalry Regiment.

1898 There were seventy members of the Twenty-fourth Infantry who provided assistance to ill yellow fever patients in Cuba.

Marksmanship Competition

1903 In 1903, seven years after the 1896 *Plessy v. Ferguson* ruling. "Separate But Equal Doctrine", somehow the white majority military authorities permitted blacks to participate in marksmanship competitions. Members of the Twenty-fourth Infantry were able to compete against white soldiers in the army wide marksmanship competition. Quartermaster Sergeant Emment Hawkins recorded the highest score in a national rifle match. "A total of 1,000 and 43 out of a possible 50".

Pursuit of Rebels in the Philippines

1906   Members of the Twenty-fourth Infantry Regiment were tasked in August 1906 to assist the Philippine constabulary in the pursuit of some rebels in Leyte.

Houston Riots

1917   The "Real Buffalo Soldiers" experienced some racial hatred, discrimination and racial violence during their tours of duty at some Forts and the nearby towns and cities. A racial incident in 1917 gained national attention and a serious concern for African Americans who were still faced with racial segregation and "second class citizenship". The Houston Texas Riot of 1917 occurred on July 29, 1917. The 3rd battalion of the Twenty-fourth Infantry Regiment was stationed in Houston to perform some special guard duties. When the Twenty-fourth Infantry arrived in Houston they were "insulted deliberately by the white inhabitants of the city. Saloons and taverns had signs reading "Dogs and Nigras not allowed here". The 3rd battalion of the Twenty-fourth Infantry had been stationed at Colona Dublan, New Mexico and were treated with kindness by the Mexicans. Later they were transferred to Columbus, New Mexico.

One evening in July, 1917, a first sergeant of Company A and some other soldiers were present in the town and decided they would not tolerate any more insults. Some confrontations occurred between the soldiers and the white citizens. The situation appeared to develop into a riot when people were wounded and the Sheriff and some people killed. There were also two soldiers killed. The Governor of Texas declared martial law and ordered the Texas National Guard to assist the local law enforcement officers in maintaining law and order in Houston. The first sergeant who led the men into town committed suicide when he "sat down on a railroad track and decided to kill himself. There was a shortage of officers in the battalion and also noncommissioned officers who were in a training school at Fort Des Moines, Iowa. It is believed that officer leadership in the battalion could have been a contributing factor in the events that occurred. An investigation revealed that the first sergeant led 150 soldiers into town and that they began to fire on the white people at random killing and wounding some. An investigation officer stated that the soldiers had defied their officers and obtained their weapons with ammunition, and departed for Houston. The investigation report stated that the white officers "showed gross negligence and insufficient leadership". Two officers were recommended to be court martialed for the violation of an article for gross neglect of duty. The commanding officer of the company involved in the race riot was Lieutenant James. He was very upset over the incidents and considered himself a disgrace. The night after he accompanied his men back to Columbus, Mexico, he returned to his living quarters and committed suicide rather than face the disgrace of having" commanded a rebellious unit.

A battalion of the white Sixty-fourth Infantry Regiment was ordered to guard the Twenty-fourth Infantry Regiment's camp site. Later they obtained a train and escorted the black soldiers back to Columbus, Mexico. Sixty three men were charged, 169 witnesses appeared for the prosecution and 27 witnesses for the defense. Thirteen black soldiers, were executed on December 11, 1917.

The black communities throughout the country demonstrated their support for the accused soldiers of the Twenty-fourth Infantry Regiment's 3rd battalion. The NAACP and other leaders requested some consideration be given for these soldiers. Numerous signature petitions were forwarded to the Department of the Army. The President, Chamber of Commerce, Columbia, Mexico sent a telegram to military authorities saying

"During the 5-6 months the Twenty-fourth Infantry Regiment soldiers were stationed here, we learned to trust and believe in them. We will be glad to have them return and other colored troops sent here". However, the Secretary of War, Newton Baker and other high ranking officials were determined to see that these men would be tried and convicted.

I sincerely believe that the killings and wounding of people during the Houston Riots of 1917 was wrong and should not have occurred. I also believe that the encouragement, enforcement and perpetuation of racial segregation and use of verbal abuses against those Black defenders of America was wrong. I also ask the question, how many white citizens of Houston, Texas who practiced segregation against the soldiers in 1917 were actually descendants of those early white settlers who the predecessors of the Twenty-fourth Infantry Regiment made it possible in many cases for the survival and future improved living condition for their ancestors? The Real Buffalo Soldiers of the Twenty-fourth Infantry Regiment paid their dues so proudly, so efficiently and so successfully even under the adverse conditions of racial hatred and sometimes injustice by the America the Beautiful they defended since 1869.

CHAPTER 4

TWENTY-FIFTH INFANTRY REGIMENT

Twenty-fifth Infantry Regiment

The 1866 Act involving the reorganization of the Regular United States Army included a provision for four black infantry regiments. However, in 1869, they were reduced to two. The Twenty-fifth Infantry Regiment were created by the consolidation of the 39th Infantry Regiment with the 40th Infantry Regiment. The Twenty-fifth Infantry Regiment was split and its battalions carried their unit designations and were attached to various divisions to replace inactive or unfilled positions. This occurred in the late 1940's.

Where Were They?

The Real Buffalo Soldiers of the Twenty-fifth Infantry Regiment were present in the "Winning of the West", and Spanish American War. Unfortunately the following historical facts are still missing from many secondary and college textbooks.

Soldiers ill with Cholera

1867 On July 9, 1867, a detachment of 25 soldiers of the Twenty-fifth Infantry Regiment were traveling to New Mexico. They camped at Fort Dodge. It was learned that some of the men were ill with cholera after being exposed to it in New Orleans, Louisiana. An officer at Fort Dodge, Major Douglas also became ill with cholera. His wife had volunteered to assist in nursing the black soldiers. Her husband survived the disease but she died.

Infantrymen Construct Buildings

1870 A company of the Twenty-fifth Infantry was responsible for the building of some structures at Fort Concho and Griffin. These talented Buffalo Soldiers demonstrated their competence in various diverse duties they were assigned. The unfortunate thing is that they could leave the military and would not have the opportunities to utilize the training they had received because of "the way it was" in those days of separate but equal life styles.

60 The Real Buffalo Soldiers

A Reconnaissance Mission

1871 In October 1871, a company of the Twenty-fifth Infantry Regiment stationed at Fort Davis, Texas accompanied two troops of the Ninth cavalry on a recon mission to the Big Bend area.

Missions and Duties

1871-1872 During the period 1871-1872, the Twenty-fifth Infantry while stationed in Texas performed various duties and missions. They were: escort duties involving long marches, escort duty, guard duty at stage and mail stations such as "Centralia, El Muerto, Barilla Springs, and different mail routes in Texas. The Twenty-fifth Regiment was also involved in road building. I can recall my childhood days observing the western movies and seeing the U.S. cavalry regiments and infantrymen riding along side the westward bound two mule buck board mail wagons and covered wagons for passengers. However, they did not tell me that some of those soldiers possessed a greater amount of melanin in their skin. Some people called them colored, black and even brunettes. But the Plains Indian called them the Buffalo Soldiers. I presume Hollywood script writers were color blind and all they could see was the all white military "winning the West for America. But in 1994, we must be reeducated to know that "They were There", Yes the Twenty-fifth Infantry "Real Buffalo Soldiers".

Escort Mission

1875 The soldiers of the Twenty-fifth Infantry Regiment experienced long and challenging marches across the far western plains. Company C, Twenty-fifth Infantry escorted a wagon train from Fort Sill, Indian territory (later the state of Oklahoma) on March 26, 1875. The party arrived at Fort Stockton, Texas after a 490 mile trip.

Expedition To Mexico

1878 June, 1878, Captain Bentzone and his Company B accompanied Colonel MacKenzie's expedition to Mexico.

Regiment Assigned To Dakotas

1880 In April 1880, the Twenty-fifth Infantry Regiment was assigned to the Department of the Dakota, at Fort Randall. Later, the regiment was transferred to Fort Snelling, Minnesota. Four companies were stationed at Fort Meade, South Dakota and two companies at Fort Hale.

1880 When Company B reported to Fort Randall, they had completed duty along the Texas and Mexican borders. They had participated in the battles of Beecher Island and Wounded Knee.

1880 Some members of the Twenty-fifth Infantry Regiment were detailed to accompany the Tenth U.S. Cavalry Regiment in the expedition that was searching for Victoria and his Indian warriors.

Guard Duty - Wagon Train

1880 When Colonel Grierson, Commanding officer of the Tenth Cavalry Regiment along with his companies D,E,F,K and L were on a reconnaissance mission, a detachment of the Twenty-fifth Infantry had a mission to guard the supply trains.

Twenty-fifth Infantry Regiment Provides Disaster Relief

1881 Some companies of the regiment were sent from Fort Randall to provide disaster relief for the white settlers in the Dakota Territory. These soldiers of color who were faced with racial discrimination and oppressions because of their skin color would always execute their orders and commands in an outstanding manner.

These Buffalo Soldiers of the Twenty-fifth Regiment provided disaster relief for the white settlers, many immigrants in the Dakota Territory. Those concerned soldiers assisted some 800 men and women. They also gave protection for the livestock.

Railroad Guard Duty

1881 Although blacks were discriminated aboard passenger railway trains in the United States in 1881, people of color were providing protection for those railway trains. In May 1881, the Twenty-fifth Infantry Regiment along with the Seventh Cavalry Regiment assisted in guarding the Northern Pacific Railway. Blacks did play a role in the opening of the west to the railroad lines.

Protecting The White Settlers

1882 In 1882 some companies left Fort Randall to provide protection for the white settlers along the Keya Paha and Montana Rivers.

Twenty-fifth Regiment Transfers to Montana

1888 The Twenty-fifth Infantry Regiment exchanged stations with the Third Infantry Regiment at Fort Missoula, Montana, Forts Shaw and Custer.

Assistance To The Settlers

1890 Some soldiers of the Twenty-fifth Infantry Regiment were dispatched to assist the white settlers living north of Flat Head in north western Montana. They also protected the settlers from a Canadian Indian tribe living on a nearby reservation.

Twenty-fourth Soldiers Maintain Order In Idaho

1891 In 1892, there were labor disputes between mine owners and labor unions in the Couer D. Alene Mining District of Idaho. On July 4, 1892, an open war occurred. There was a serious riot causing some mines to be destroyed and some loss of life. The federal government decided to intervene on July 12. The military authorities were directed to send Federal troops to stop the unrest. Among the troops were a provisional battalion of the Twenty-fifth Infantry Regiment. The regiment was ordered from Fort Missoula to join other army troops in maintaining law and order at the mining companies. The Buffalo Soldiers of Companies F,G,H, 148 enlisted men were sent to the Idaho Mining District, Millan, Idaho. Some years ago, a Colonel, U.S. Army, who happened to be a Director of Intelligence at the time, asked me did they have "black soldiers in the army prior to World War I. I guess they did not tell him in the military classes where he completed his upward mobility strides in attaining the rank of colonel about the Buffalo Soldiers. He was from the western states, possibly Idaho.

Escort Duty For The Railroad

1892   Around July 13, 1892, a Northern Pacific Railroad track was destroyed in two areas. Some soldiers of the Twenty-fifth Infantry Regiment were dispatched to provide security, and assist in making arrests.

A Search Party

1893   There were some hunters in 1893 known as the Carlin group who were snow bound in the mountains. The Buffalo Soldiers of the Twenty-fifth Infantry Regiment marched a distance of ninety miles. They were not successful in locating the stranded party.

1894   When Jacob Coxey became concerned about mass unemployment he suggested a "500 million dollar government public works program. He believed the program should have been funded with paper money not backed by gold, but just use legal tender. Coxey had many followers. He organized a march on Washington in March 1894. Jacob Coxey also had organized groups in other cities. On May 6, 1894, some members of the Coxey Army in Montana were arrested. The U.S. Marshall had requested some military assistance. Members of the Twenty-fifth Infantry Regiment were sent to assist the Marshall in controlling the Coxey followers at Arlie. The Buffalo Soldiers of the Twenty-fifth Regiment returned to Fort Missoula on June 3, 1894.

Guarding Railroads - Montana

1894   When the Northern Pacific Railroad employees were involved in a serious disagreement with management, there were fears of strikes, riots and disorder. The concern for the mail and passenger trains caused officials to request government assistance. There was a need to protect the mail, workshop facilities of the railroad, bridges, trestles and tunnels.

On July 8, 1894, the Buffalo Soldiers of the Twenty-fifth Regiment's Companies A,C and D were issued deployment orders. The companies had orders to march 34 miles to safeguard railroad property, keep safe movement of the passengers trains in Montana and protect the areas of Big Timber, and Miner, Montana. The companies guarded railroad trains traveling from Livingston to Cokesdale, Montana. The Twenty-fifth Regiment also had guards posted near Okeefe Canyon, Montana. In some towns where they were stationed, the Buffalo Soldiers received acknowledgments of their accomplishments.

The <u>Anaconda Standard</u>, Missoula, Montana, August 6, 1924 published this account about the Twenty-fifth Infantry Regiment.

"The men acted with wisdom under Captain Anderson while guarding Marent and Okeefe's trestles and depots yards. Some prejudices did exist. However, the model soldiers conduct was excellent and they were orderly and quiet".

Bravery In Cuba, 1898

1898 On July 1, 1898, Buffalo Soldiers of the Twenty-fifth Infantry Regiment were assigned a mission of advancing under heavy enemy fire to secure a stone blockhouse. The members of the Twenty-fifth Regiment Infantry displayed courage and aggressiveness in accomplishing their mission.

Bravery in the Philippines

1899 During the Philippine Insurrection, the Twenty-fifth Infantry Regiment attacked the town of O'Donnel on Luzon and captured 128 prisoners and rifles with ammunition on November 17-18, 1899.

Twenty-fifth Soldiers On Luzon

1899 In 1900 , on the island of Luzon, Filipino insurgents were present in many provinces. The Twenty-fifth infantry Regiment executed a surprise attack and captured a strong hold in the town of Mayalang, province of Pambanga. The province was an important strategic position for the insurgents. The rebels were located on the top of a high hill where they were able to protect themselves. Eventually, the Buffalo Soldiers of the Twenty-fifth Infantry Regiment were able to capture the rebels.

A Skirmish at San Mateo

1899 On August 12, 1899, 150 men of the Twenty-fifth Infantry Regiment were involved in a skirmish between some Filipino rebels in San Mateo, Philippines.

Brownsville Affair

1906 Until September 1972, official and public accounts of the Brownsville Affair had been summarized as follows:

On August 13, 1906, three companies of the all-black companies (except officers) were involved in a riot in Brownsville, Texas. The white citizens of the town had alleged that the blacks had shot up the town and that Negroes had murdered and maimed the (white) citizens of Brownsville. One citizen was killed, one wounded and the Chief of Police was injured. Based on official military reports and other investigations, the President of the United States in 1906, Theodore Roosevelt, of Spanish-American war fame, dismissed the entire battalion from federal military service without honor and disqualified the members from future enlistment and loss of benefits.

Many Americans both black and white protested. John Miholland, of the Constitution League rallied in the support of the dismissed soldiers. Senator Joseph B. Foraker, Ohio, demanded a full and fair trial for the soldiers. The senate responded to his request by authorizing a general investigation. After several months, the majority members of the senate committee upheld the President's decision.

September 28, 1972, Secretary of the Army Robert F. Froehlke announced that an army review board had reviewed the 1906 discharges of 167 black soldiers of the First Battalion, Twenty-fifth Infantry (Colored) who were discharged in 1906 without honor as a result of a shooting incident which occurred in Brownsville, Texas.

The army's summary of the incident was: "Around midnight on August 13, 1906, some sixteen to twenty individuals on horseback rode through the streets of Brownsville firing their weapons into homes and stores. As a result of the shooting, one man was killed and two were injured. Witnesses alleged that the raiders were colored soldiers. At this time the First Battalion, Twenty-fifth Infantry (Colored) was stationed outside of the town of Brownsville.

A series of military inquiries and a county grand jury failed to establish the identity of the riders involved. Finally all members of companies B, C, D of the First Battalion were assembled and the guilty were told to step forward and identify themselves or all would be discharged without honor. None stepped forward; all maintained their innocence. Their discharge without honor followed. Subsequent courts of inquiry failed to recommend remedial action and relief legislation introduced on behalf of various individuals was never enacted. An internal army review of administrative and judicial policies brought this instance of mass punishment to the secretary. Although the practice was occasionally invoked under extreme circumstances during frontier times, the concept of mass punishment has for decades been contrary to army policy and is considered gross injustice.

The results of the military review of the Brownsville Affair enabled Dorsie Willis, eight-seven years of age, Minneapolis, Minnesota, and Edward Warfield of Los Angeles, possibly the only survivors of the group of the Twenty-fifth Infantry unit that received dishonorable discharges, to be exonerated.

Even though the military new release stated that "the practice (decision concerning 167 black soldiers in 1906) was occasionally invoked under extreme circumstances during frontier times," the known facts today, and those of 1906 pose two questions. Was it a normal practice in an area not actually a frontier, Brownsville, Texas, to dismiss 167 men from the military service dishonorably, with the final decision executed by the President of the United States? Could the Brownsville affair's decision of mass punishment have been politically oriented? A noted black historian, Professor Rayford W. Logan stated, "One must conjecture whether Roosevelt's [Theodore Roosevelt] abrupt order for the dishonorable discharge of three companies of the Twenty-fifth Infantry after the Brownsville Riot stemmed from a further desire to propitiate the south.

Fighting Forest Fires in the South.

1910 The Buffalo Soldiers of the Twenty-fifth Infantry Regiment were detailed in the summer of 1910 to assist in fighting forest fires that had occurred in the Glacier National Park in Montana. A company under the command of Lt. W.S. Mapes was given the task. Mapes company was stationed at Essex Montana. A company under the command of Sergeant John James was assigned the job of supervising and protecting the people during the evacuation from the area. James' company was stationed at Fort George Wright Washington, Another company was commanded by Lieutenant E. Lewis. His company had the mission of patrolling and providing security for the railroad trains.

The U.S. Supreme Court had ruled around 1900 that federal troops could be used in states to suppress riots and maintain law and order without the consent of the state. The Twenty-fifth Infantry Regiment was used in some instances to provide assistance and maintain law and order when ordered by the federal government. In later years, volunteer state militia and/or National Guards units would be deployed instead of regular army units. One must remember that state militia in the majority of the states in 1900 had few if any black members in their units. Segregation was the rule. But somehow during the days of legally enforced segregation, the United States Army used soldiers of color to confront white citizens when they were involved in riots or civil unrest. Yes, the Buffalo Soldiers were there on the scene.

CHAPTER 5

SEMINOLE NEGRO INDIAN SCOUTS

Seminole Negro Indian Scouts

The Black Indian Fighters

In America's conquest and defense of her western and southwestern frontiers black men with Indian blood, the Seminole Negro Indian scouts, played an integral role. During the days of slavery, free Negroes and runaway slaves needed havens of freedom and tranquility. Some went to far northern states and Canada and small groups found their way into Indian camps where some became slaves again and others intermarried and cohabited with the Indian tribes. The Seminole Indians of Florida saw the Negro come and go, but while Negroes lived with the Indians they miscegenated and the two races became mixed. American Negroes can claim an Indian genetic heritage which in some cases is physically dominant.

In Texas today there are descendants of a small group of Negroes who joined the Seminole Indians some 120 years ago. Many stories have been told concerning their ancestors, but little has been written. The following paragraphs attempt a brief description of this group of black defenders whose achievements, although they include four Medal of Honor awards to Seminole Negro Indian scouts, are not widely known.

The Seminole Negro Indian spoke both an Indian language and English. They were mostly Baptists, but in many cases observed Indian ritual and custom. Their homes were in separate villages where they tended their own fields and cattle. A chief, Negro Chief John Horse, had been chosen among the group and in 1849, with Chief Wild Cat, he led a group of dissenting Seminole Indians who departed for Mexico. They eventually settled in the area of Laguna de Parras, southwestern Coahuila, around 1870. Other Negro-Indian groups had settled at Nacimiento, Coahuila, Matamoras and on the Nueces River, according to official correspondence between the post at Fort Clark, Texas, and the adjutant general, Department of Texas, San Antonio, Texas, dated August 26, 1884.

Captain Perry of the Twenty-fourth Infantry Regiment went to Mexico and asked the Seminole Negro Indians to come on the Texas side of the Rio Grande and the government would give them a reservation. It was also stated that General McKenzie told the group that had arrived on the Texas side and become scouts not to return in 1879 because they would later be provided with a reservation. Those scouts that had departed Mexico were supplied with arms, ammunition and rations. They provided their own horses and received compensation. Their dress was similar to Indian style.

Over a period of thirteen to fifteen years the Negro Indian scouts performed their missions in a superb manner. Their contributions to the regular army units were commendable and essential. The part they played, though it may have appeared insignificant, was instrumental in achieving many victories and successes for the regular army units they were assisting. Some examples of their heroic action follow.

In 1874, Colonel MacKenzie was on an expedition against the Cheyenne Comanche and Kiowa Indians in Palo Duro Canyon. During this expedition twenty-one Seminole Negro scouts accompanied him and displayed unusual bravery. . . . The scouts searched for horse thieves in Mexico and hunted Indian raiders. In many cases they were utilized to seek out the location of hostile enemy camps.

The Seminole Negro Indian Scouts and their families resided at Forts Duncan and Clark. In 1873, the Indian scouts realized that they could not forever exist on government reservations, therefore they began to demand the responsive action to the alleged promises they had received. There was considerable discussion as to the disposition of the scouts and their families. Correspondence between the Executive Office, Seminole Nation and the U.S. Indian agent, dated September 17, 1883, stated that the Seminole Nation cannot recognize the Negro Scouts' claim to return to Seminole reservations and that because of their flight to Mexico, they have no claim. The principal chief [of the] Seminoles, John Jumper was quite adamant in his refusal to accept the Negro Indian scouts. Official correspondence around revealed that the government was indecisive as to where the Negro Indian scouts and their families should be located.

In consideration of their devotion to duty and country, their earnest fidelity and loyalty as scouts, it was believed by some supporters that they should be given adequate land to satisfy their existing problems.

Excerpts from correspondence written by two officers of the United States Army who knew the Indian scouts and were aware of their tribulations should best illustrate the urgency and importance of their request during that time.

They have undoubtedly lost all claim to their lands given them by the Mexican government, and it is not known whether that government will see fit to renew the grant, but in any event, I believe it would be cheaper to place them on a United States Reservation. They have done good service as scouts during the last thirteen years, and are entitled to consideration. I think if a reservation was given them, nearly all would go to it, if not at once, within a very short period. Some of them ask permission to occupy the houses they have built on the reservation until the spring of 1885, when they hope if not otherwise provided for to be able to obtain work in Texas, and support their families, some of them I think will return to Mexico in any event.

Z.R. Bliss, Lieutenant Colonel, Nineteenth Infantry, Commanding Post, Fort Clark, Texas, 26 August, 1884.

I now forward the enclosed correspondence to show how hopeless it is to expect any relief from the Indian Bureau. Thirty-four men all with wives and children who have served as soldiers for the average of thirteen years, without any regards of property and with habits essentially Indian, are thrown upon a community itself poor and hostile to these harmless vagabonds.

The thirty-four enlisted scouts to be discharged represent at the least one hundred and fifty souls, and how they are to live, or what is to become of them, I cannot imagine.

To turn them loose upon the people of Kinney County, wherein they now are, is hardly a right thing to do and will probably lead to trouble. My impression is that they should be sent to the Indian Territory and settled in the manner of the Modoes and Nez Perces.

D.H. Stanley, Brigadier General, Commanding, Headquarters Department of Texas, San Antonio August 27, 1884.

The proud and loyal Negro Seminole Indian scouts were disbanded around 1914. These dedicated soldiers and their families were not granted all their requests for land and security, but they have attained a place of distinction and honor in the history of America's black defenders.

Ref. *Black Defenders of America 1775-1973.*

CHAPTER 6

THE PLAIN INDIANS

The Plain Indians

The Buffalo Soldiers who served during the Indian Campaigns met the Plains Indian warriors in brief and extended skirmishes and battles. I believe it is appropriate to present some major characteristics and lifestyles of the Plains Indians. Unfortunately, through the years the Native American has been portrayed by cinema, authors and play writers in a negative and stereotype manner As late as July 1994, one can view white actors and actresses still playing the part of Indians. There were many mixed whites among the Indians due to miscegenation between the Indian and whites. However, we do not hear the term "mixed blood or half breeds" for Indians who miscegenated with blacks. The Smithsonian Institute Bureau of American Ethnology has a very valuable collection of photographs depicting the Native American. An examination of these photographs and a careful observation of the distinct physical characteristics of these Indians, especially skin color, one can conclude that the skin complexion of some Indian tribes were of the distinct colors very dark brown to black using the American racial complexion classification of people. I observed those shades of complexion of some Indians representative of the Cheyenne, Kiowa Comanche, and Apache. Their complexion certainly did not represent the skin color depicted as Indians in movies and television presentations. Even the complexion description could have confused enrollment officers when signing up Indian scouts. An enlistment officer in Arizona in 1884 listed the physical description of the Indian scouts he enrolled. The physical description included height, in feet, inches, complexion, and color of eyes and hair. The descriptive terms used for complexion were "light, copper and black". Hair was described as black, and black and curly. It should be noted that the descriptive rolls of registration for African American or black in many cases reflected the same descriptive characteristics, especially the rolls of the Seminole Negro Indian scouts in Texas. I recorded the following information from the descriptive rolls of the Arizona Warm Springs and Chiricahua Apache Indian scouts. These descriptions reflect only what the caucasian officers used to describe the Indians they enrolled. In no way am I stating that these Indians were black. However, there was and still is in America a problem of color in classifying people not by who they really are but as the visible eye so dictates. These Indians were given their physical description by the white enrolling officers.

Name	Age	Height	Complexion	Color Eyes	Color Hair
Benito, Chief	45	5'8	copper	black	black
Cheio	60	5'5	black	black	black
Shore	47	5'7	black	black	black
Qui-hoi-inney	22	5'7	black	black	dark
Cath-h	35	5'8	light	black	black and curly
Om-my	20	5'3	black	black	black
Dutchey	25	5'4	copper	black	curly and black

    I infer that the many stories concerning how the Buffalo Soldiers received their name from the Indians, certainly could not have had reference to their skin color because the Indian warriors they were facing on the battlefield had one thing in common with them and that was the different shades of skin color from light-copper-black. A German saying I learned some 30 years ago "Die Farbe ist egal" (the color makes no difference) possibly also applied to the Indian warrior when they saw the people of color in the distinctive uniform of the U.S. Cavalry and Infantry regiments to include part of their genetic blood line, the Seminole Negro Indian Scout.

    Who were the Plains Indians? They were a proud people, unified by tribal connections, who cherished their traditions and culture. They believed that they were right in resisting the white man who had removed their buffalo through almost extermination and took their land through valid and invalid treaties and finally drilled them to a reservation without rights as a citizen of these new United States of America. The Native Indian witnessed the occupation and claiming of lands rich with minerals, gold deposits, rich soils and forests with valuable tree resources. The Indian campaigns were a resistance movement for the Indians against the white man's exploitation and conquering a rich and prosperous Far Western Frontier for America the Beautiful".

During the Indian Wars, there were some thirty Indian tribes that lived in the Great Plains area. Some of the major rebel groups that fought skirmishes and battles with the Buffalo Soldiers were: the Sioux, (Lakota and Dakota), Comanche, Cheyenne, Arapahos, Kickapoos, and Kiowas.

The Great Plains area covered approximately 780,000 square miles, from the Saskatchewan River to the North Rio Grande to the south and to the Rocky Mountains to the west. Historically, the North American Indians came from north eastern Asia crossing the Bering Land Strait many years ago. The North American Indians spoke six different languages. They were Algonquian, Athapaskan, Caddoan, Iowan, Sioux and Uto-Aztecan had various dialects. The Plains Indian physical description was "slim statue, 5' 10", small hands and feet, brown eyes, straight black hair little facial or body hair and skin color or complexion from light brown to dark brown to black."

The Plains Indian were a nomadic group of people who traveled their vast land areas and moved from place to place often to obtain food and sufficient shelter for survival. The Plains buffalo played a very significant part in the life of the Indians. The Indians hunted the buffalo and used the animal for many purposes. The horns were used to make cups, spoons, head dress and toys for the children. The hair was used to fashion ropes, head dress and medicine balls. The Indian's outstanding creative abilities were successful in using the buffalo' bones to make knives, shovels and the animal's hide for buck skin cloth, gun cases and tepee cover. The tail was used to make rawhide, shields saddles and shoe ropes. The bladder of the buffalo was used to make water bags.

The Plains Indians family life was centered around a structured society consisting of a "tribal chief, medicine man, warriors, craftsmen, clan chiefs, band chiefs and the Family Band." The Indians of the Plain utilized the natural reserve of animal available for hunting. They were able to hunt the buffalo, deer, antelope, elk moose, wolves, coyotes, bobcats and they also had available the rabbits, squirrels, foxes, otters, badgers, beavers, chicken, and wild turkey. The arrival of the white man and his determined desire to exploit the western frontier and eventually settle in the Plains area definitely contributed to the unfortunate removal of the Plains Indians from their home lands and productive food supplies and other natural material that were essential for their existence.

The Plain Indians' diet consisted of buffalo meat, wild berries, strawberry, plums, marshmallow, peppermint, turnips, melons, pumpkins cabbage, beans, corn, onions, wheat, peppers and barley.

Some characteristics and lifestyles of the Plains Indians are: The Plains Indians believed their most precious gift was their children. The Plains Indians performed a Sun Dance. The dance included materials such as a sacred cottonwood tree, streaming flags of colored cloth, small tobacco bundles, and pin branches that would form a roof over a circle of poles. The dancers would form a line and pray. Sometimes the dance would last for several days. The Plains Indians had sacred bundles made of cloth and contained objects they believed had magical powers. The warriors would carry them into battles. Young Indian boys played a game called shinny which later became the game ice hockey. The Indians would use a long wooden stick to knock a ball over a goal line. The ball was made of baked clay covered with buckskin (soft deer skin).

The Indian warrior was trained to fight at a young age. The Sioux warriors used rifles, bows and arrows. The Indians conducted raids in many instances to obtain food and livestock. Their raiding parties consisted of 5-15 men. Their movements would be a slow pace and they would travel early in the morning. Some Cheyenne warriors were called "Dog Soldiers". These Indians would go into battle and a party of ten men would ride ahead of the main party. One man would dismount then drive a lance through a black sash tied around his neck. Then he would remain on the ground as the battle was going on and he believed this action would provide support to his comrades. A high honor for the Plains Indians was to touch a special weapon, landing a direct blow. The Cheyenne Tribe was the first tribe to use the rope sling around the body of their horses.

The Plains Indians were outstanding horsemen. They were able to slip down on the side of the horse's body with their heel of one foot placed in the horse's back. The weight of their bodies was secured by a rope. They would appear to one approaching that it was a riderless, horse. They could also place their hands on the horse's backside and vault onto the horse's back from the rear, this was mostly executed on ponies. The arrow of a Cheyenne Indian could travel a distance of 165 yards. The arrow had grooves cut in the shafts. This provided the straightness of the arrow when it was in flight and upon penetration into the flesh of the enemy, the grooves facilitated an increase in the flow of blood from the person's wound.

The Plains Indians would wear feathers according to things they had achieved. The Apaches would use war paint on their chest and hands. The colors were black and yellow. Scalping was practiced by some Indians and Mexicans. Some Indians used methods of torture against their enemies. They would also take captives and sometimes adopt them into their tribes. Scalping did not originate directly from the Indians. It is believed that the early British settlers would scalp their enemy to prove they had killed an Indian. They brought back the scalp and were paid in goods, weapon and rum. The British also had a bounty of forty English pounds in 1755 for the scalp of Native American that were obtained from the unfriendly tribes who were fighting other Indian Tribes who had probably made peace with the British. The term Apache was given to the Indians by a tribe of Arizona that was their enemy. This tribe called them the Apaches and later the white man changed it to Apache.

Prior to the 1900's the Indians had been driven to the reservations by the white majority Americans. The reservations were land allotted for Indians to live on by the United States government. Some of the major reservations are: Fort Apache, Camp Verde, Camp Grant, Ojo Caliente, Pine Ridge, San Carlos, Mescalero and White Mountain.

Some great and proud Indian leaders that engaged in skirmishes and battles with the Buffalo Soldiers were Victoria, Geronimo, Roman Nose, Satank, Big Tree, Nana, and Mangas.

The United States Army recruited some friendly Indians from the various tribes to serve as Indian scouts. They were representatives of the White Mountain, Apache, Cheyenne, and Navaho. Indian scouts received the Congressional Medal of Honor also Seminole Negro Indian scouts.

The Native American has been recognized in different ways by American society. Some have been laudatory. There is a 1875 U.S. Twenty dollar bill with the appearance of a woman resembling Pocahontas. The U.S. Mint in 1913 issued a buffalo head nickel, one side displayed the buffalo, the other side an Indian Chief. Three Indian Tribes were used in composing the portrait of the Indians, they were the Cheyenne, Sioux and Seneca. The Plain Indians lived and hunted in territories that later would become states that beared an Indian name. The states are Arizona, Kansas, Idaho, Nebraska, New Mexico, Oklahoma, Minnesota, North Dakota, South Dakota, Texas, Utah, Wisconsin and Wyoming.

The Native American today is still confronting some vestiges of injustice, denials and respect as the Native people of this great country. As a minority the American Indian also realizes as African Americans do, that all equal rights and all equality in America is a continuing challenge to a country that must provide these right for all Americans.

CHAPTER 7

INDIANS WARS

Chapter 7

Indian Wars

The Indian Wars of the United States around 1866 and 1890 were a series of skirmishes fought in the western and south western United States. Regular army soldiers and scouts were involved and African Americans participated in both capacities. The aim of these wars or campaigns was to protect the interest of (mainly white) settlers who had moved into Indian territory.

In 1866, the Ninth and Tenth Cavalry regiment after their organization began their duties at various western posts. The Twenty-fourth and Twenty-fifth Infantry Regiments in 1869 commenced their specific duties on the western frontier. These soldiers of color, the "Real Buffalo Soldiers" were present in campaigns against the Apaches Kiowas, Cheyennes, Comanches, and Arapahos. They escorted railroad surveyors, settlers, protected mail and stage coach between San Antonio and El Paso, Texas. These regiments of color were active in the Victoria War and the pursuit of Indian Chief Geronimo.

During the Indian Wars, Indian scouts were used by the military. The black presence in the Indian Wars merited the awarding of 14 medal of honors to the members of the four brave regiments of color and four medals to the Seminole Negro Indian scouts.

The Buffalo regiments during the Indian wars or campaigns engaged in many skirmishes or battles with the Indian warriors who were trying to protect their own land and existence on the Great Plains. The black soldier, many recently freed from slavery, enlisted in the four regiments for many reasons. The reality and the facts are that men of color who had a historical legacy and in some instances personal experiences of being a former slave were , members of the United States regular army cavalry and infantry units in 1866 and 1869. Some of them had performed in an excellent manner as a Union soldier in the Civil War. Their personal desires, initiative, discipline, courage and self concept and determination all contributed to the outstanding performance that were accomplished by the majority of those men who had severed around 1867-1890. Their combat records or engagements and fighting the Indian warriors proved beyond any doubt that people of African descent to include those biological genetically diverse with other racial groups especially Caucasian and Indians were able to perform as good as and in some cases exceptional. The argument that they performed well under the leadership and presence of the white officer always will be debatable. Because the records of performances of some noncommissioned officers in the absence of their white officers proved those Buffalo Soldiers of the Frontier West were capable to lead and make decisions. The men of the four brave black regiments were involved and present in the following events during the "Winning of the West".

1868   Engaged in skirmishes with the Indians at Sandy Creek.

Soldiers of Tenth Cavalry present at Beecher Island.

During Sherman's Campaign, black troops participated in the pursuit of Indian warriors with legendary Wild Bill Hickok.

1869   Buffalo Soldiers confronted Indian warriors at Kickapoo area near, Ft. McKavett.

1871   Troops present at Camp Supply, Indian territory patrolled along the Red River to prevent raids by Indians.

1874-
1875   Black soldiers participated in the Red River War.

1875   While escorting some Indians to a Florida prison, several encounters occurred with some Indian warriors. The soldiers were able to force them to flee.

Buffalo soldiers were present with an expedition force that was mapping parts of the country for white settlers in Texas and eastern Mexico.

1876   The Ninth U.S. Cavalry Soldiers were trying to locate some Apache warriors in New Mexico.

1877   The Tenth U.S. Cavalry soldiers were chasing Black Horse and his Indian warriors when they left the Fort Sill, Indian reservation.

1880s   The army's pursuit of Victoria and his Indian warriors included the participation of the Buffalo soldiers.

1890   The Ninth Cavalry Regiment was present at Wounded Knee to provide rescue assistance to Seventh U.S. Cavalry Regiment.

In a mystic way, the current U.S. Postal Buffalo Soldier stamp becomes a "crystal ball" and one can see the presence of those "Real Buffalo soldiers" at:

Saline River	Red River War
Sandy Creek	Palo Duro Canyon
Camp Supply Indian territory	Saragossa
Beaver Creek	Pinto Mountain
Beecher Island	Pecos River
Wichita	Sierra Canned
Buffalo Springs	Shakehand Springs
Sulphur Springs	Rattlesnake Springs
Tinaja De Las Palmas	Ojo Caliente
Eagle Springs	White mountains
Alamo Springs	Wounded Knee (Drexel Mission)

"Buffalo Soldier" origin of a name

Throughout the years, legends, writers and authors all have given different views and versions of how the soldiers of the four black regiments in the West received their name "Buffalo soldiers". People have said:

"Native Americans used the name because of the soldiers's bravery and courage on the battlefield".

"The Indians, observed the black troopers wooly hair that resembled the shaggy coat of the sacred Buffalo".

"The black soldiers heads are matted like the hair between the horns of the buffalo"

A Buffalo Soldier's Remembrance of the Indian Campaigns

Sergeant Richard Anderson, Ninth U.S. Cavalry Regiment wrote the following in 1898. "While stationed in the sub Camp at Fort Cummings, New Mexico in 1880, I received orders for campaign duty against Chief Nana and his band of Indian warriors. On June 5, 1880, my troop commander was absent from Ft. Bayard and I was left in command of Troop B. A band of Apache Indians were marching toward Cooks' Canon. At this time, Troop B and I under the command of Captain Frances, Ninth Cavalry were ordered to leave for our mission to pursue the raiding Indian bands. We arrived at Cooks' Canon and observed the Indians in the area. There was a skirmish that lasted almost three hours. A few troopers were wounded. After following the Indians who retreated, a decision was made to return to Fort Cummings, New Mexico".

Sometimes the exceptional performances and bravery in combat experiences of enlisted men and noncommissioned officers are not recorded in the literature as frequently as their officers' exploits. During my literature search for this manuscript, I was able to support this inference because numerous accounts of the cavalry and infantry units' performance centered around the white officers of the regiments and companies. The outstanding writings about the men of the black regiments in the Spanish American War has been a great asset in the preservation of black performance in combat conditions. This was accomplished so eloquently by Chaplain Stewart in his book. He also recorded another account of Sergeant Anderson's remembrances of the Indian campaign.

Sergeant Richard Anderson recalled his experiences in combat in August 1881. He was ordered to go on a 30 day detached duty with his troop and Troop H. His commanding officer was absent on court marital duty at Fort Bayard. With no other officer available at that time, Anderson assumed command of Troop B. He was under the general command of Lt. Smith of Troop H. Troops B and H departed Ft. Cummings for Lake Valley, New Mexico. On August 18, 1881, the troops encountered some Indian warriors who had been raiding and killing civilians in the area. During their pursuit of the Indians, the troopers marched into an ambush where Lt. Smith was killed. After the Indians fled to the mountains, Sergeant Anderson assumed complete command of the unit. He cared for the wounded and sent Lt. Smith's remains back to Ft. Bayard, New Mexico. He then marched toward Rodman Mill with his surviving men, the dead and wounded. When he arrived at Rodman Mill on August 20, 1881, he buried his dead and sent the wounded to Fort Bayard.

One of his wounded men, Private John W. Williams, Troop H, Ninth Cavalry had been shot through the knee cap and he had to ride all night to Brookman Mill. After arriving at Fort Cummings, the regimental commander, Col. Hatch asked Sergeant Anderson could he mount up the next day for a pursuit mission. On August 21, 1881, this outstanding NCO of the Ninth Cavalry responded to orders and mounted up and ordered his troop to accompany Troop L along with Lt. Dommick to follow an Indian Trail. They followed the Indians to the border of Old Mexico" and then returned to the Fort.

It was during the Indian Campaigns that many Buffalo soldiers demonstrated their efficient and superb abilities not only to receive orders and follow but also to be prepared to give orders and lead in circumstances where their white officers would be absent. A reflection, I sincerely believe that those honorable learned senior military officers at the U.S. Army War College and higher headquarters commands around 1900-1950 could not in using sincere reasoning and logical thoughts really believe that blacks could not perform under combat conditions and should not fight in combat. I personally give them more credit for whatever intelligence they possessed in their decision. In view of the black's past presence under combat conditions in the American Revolution, War of 1812, Mexican War, Civil War, Spanish American War and the unquestionable outstanding performances during the Indian Campaigns by the Buffalo soldiers those so call honorable white architects of military policies for the U.S. military soldiers must have been guided by a central theme, enforcement of the Plessy v. Ferguson, Separate but Equal Doctrine, of 1896 in all aspects of American Life to include the military service.

Unfortunately, the world viewed the true meaning of these statements that were so vividly depicted on June 6, 1994 with the small numerical observance of Black Defenders of America in combat on D-Day of WW II's final push to contain Germany's advance. Where were we? Simply complying to America's wishes of a separate policy. Ironically, the Buffalo Soldier was present on the home front in the 1870-1880 to open the way for America's Western Frontier to white settlers. However, in the 1940's the U.S. military would continue to retain its separate black units but would integrate the American Indian and Hispanic American into the white U.S. military units. They were there on D-Day 1944. There were also racial designations. Namely Asian, black and white. I can only presume that all who were in white units, that their racial designation on official records must have been listed as "white". Those brave Buffalo soldiers of yesteryears, enlisted and non commissioned officers, have brought the people of color thus far in their pursuit of excellence in military careers. One cannot forget that the early manpower input for black officers came from the NCO rank. The first black military, general officer on active duty was a former Sergeant Major in the Tenth U.S. Cavalry, the late Brigadier General Benjamin O. Davis Sr.

The military Surgeon's treatment of wounds and injuries during the Indian War was based on available medical knowledge, equipment, surgical instruments, medicine and type of injuries incidental to method of warfare. A major injury that was frequently treated by surgeons was the arrow wound received from the Plains Indian warrior's type of arrow and body target area that was hit. The Indian Warrior would aim his arrow at the soldiers' abdominal area, aiming at the abdomen (umbilicus). The Apache Indians were using arrows in warfare as late as 1885. Some arrows were of the barbed type and the hits were highly successful. There were some soldiers who died from arrow wounds due to complications from shock, blood poisoning and a severed artery. A major procedure used by the regimental surgeon to treat arrow wounds involved using duck bill forceps to remove the arrow. The surgeon would dilate the area where the arrow entered, and attempt to remove it. Sometimes there would be crushed parts of the arrow. A Buffalo Soldier, Private Samuel Brown, a trooper of the Tenth U.S. Cavalry received a serious arrow wound in December 2, 1868 near the Canadian River, Texas. The arrow entered the abdominal area of Brown's body. It penetrated in the region of the left "hypochondria region". This injury caused a punctured wound three quarters of an inch in length. It then caused the protrusion of 18 inches of the small intestine outward. The Surgeon closed the wound of the intestine which was dissected or cut in four pieces. The surgeon used sutures to close the wound. Private Brown's physical weak condition and shock caused by the injury contributed to his death.

Epilogue of The Buffalo soldiers

The courage, daring acts of bravery and gallant military service in the Far West have been commendable. I sincerely trust that these historical facts of the black presence in the Indian War will give a serious positive image of blacks in the American experience.

CHAPTER 8

SPANISH AMERICAN WAR

Spanish American War 1898

The Spanish American War of 1898 saw the African American officer take his place as a leader of his men. The National Guard and state militias were a major part of the country's military force. Several of the black volunteer units were commanded by senior black field grade officers.

During the war, black participation was evident. There were the Ninth and Tenth Cavalry Regiments and the Twenty-fourth and Twenty fifth Infantry Regiments present in Cuba at San Juan Hill and at El Caney making history on the battlefields. The Spanish American War was fertile ground for the growth and development of a new black soldier, a commander and leader of his black troops. The non commissioned officer (NCO) of the four black regular army regiments provided a great resource for the commission of many black officers during the Spanish American War.

Spanish American War Casualties

During the Spanish American war's combat engagements in Cuba, some courageous Buffalo soldiers of the Ninth and Tenth Cavalry regiments and Twenty-fourth and Twenty-fifth Infantry Regiments were killed and wounded in battle. I have included some of the names of these men because I believe it is necessary that we recall the truth and facts that these Black Defenders of America did die in the line of combat operations for "America the Beautiful" some 86 years ago:

Ninth Cavalry Regiment Men Killed

Trumpeter Lewis Fort and Private James Johnson.

Wounded

<u>First sergeants</u> Charles W. Jefferson, Thomas B. Craig,
<u>Corporals</u> James W. Ervine, John Mason,
<u>Privates</u>, Hoyle Irvin, James Gandy, Edward Nelson, Noah Prince,
    Thomas Sinclair, James Spear, Jacob Tiell, Burwell Bullock,
    Elijah Crippen, Edward Davis, Alfred Wilson, George Warren,
    and William H. Turner

Tenth Cavalry Regiment Men Killed

<u>Corporal</u> W. F. Johnson
<u>Privates</u> John H. Smoot, John H. Dodson, George Stroal, and
   William H. Slaughter

Wounded

<u>First sergeants</u>, A. Houston and Robert Milbrown
<u>Quartermaster Sergeants</u> Smith Johnson, Ed Lane, Walker,
   George Dyers, Willis Hatcher, John L. Taylor, Amos Elliston,
   Frank Rankin, E.S. Washington, U.G. Gunter
<u>Corporals</u> J.G. Mitchell, Allen Jones, Marcellas Wright
<u>Privates</u> John Arnold, Charles Arthur, John Brown, H.W. Brown,
   William A. Cooper, John Chinn, J.H. Campbell, Henry Fern,
   Benjamin Franklin, Isom Taylor, Gilmore Givens,
   B.F. Gaskins, William Gregory, Luther D. Gould, Wiley Hipsher,
   Benjamin West, Thomas Hardy, Joseph Williams, Charles Hopkins,
   Richard James, Nathan Wyatt, Wesley Jones, Robert E. Lee,
   Sprague Lewis, Harry D. Sturgis, Peter Saunderson,
   John T. Taylor, Henry McArmack, Samuel T. Minor, John Watson,
   Lewis Marshall, Allen E. White, William Matthews,
   Houston Riddell Charles Robinson, Frank Ridgeley and
   Fred Shacker.

Acts of Bravery

In June 1898, Troop I, Tenth Cavalry Regiment commanded by Lieutenant R.J. Fleming had some of his men to distinguish themselves in battle by demonstrating "coolness" and gallantry". They were Farrier Sherman Harris, Wagoneer John Boland, and Private Elsie Jones.

The Twenty-fifth Infantry Regiment at El Caney, Cuba

Chaplain Stewart stated in his book *The Colored Regulars In The United States Army, 1903* that an editorial appeared in a religious newspaper and praised the exploits of the Twenty-fifth Infantry Regiment,

"American valor never shone with greater luster than when the Twenty-fifth Infantry Regiment swept up the sizzling hill of El Caney to the rescue of the Roosevelt's Rough Riders. But the bullets were flying like driving hail. The

enemy were in trees and ambushes with smokeless powder and the Rough Riders were biting the dusk and were threatened with annihilation".

Twenty-fifth Infantry Regiment Bravery In Combat, Cuba

The commanding officer of the Twenty-fifth Infantry Regiment Lieutenant Colonel A.S. Daggett was concerned about his regiment not receiving recognition for their commendable accomplishments in combat during the Spanish American War. Daggett wrote his higher headquarters and stated that:

"When the Twelfth Infantry in rear of the Fort (Blockhouse) completely sheltered from the enemy's fire received the white flag of the insurgents. However, Privates J.H. Jones, Company D and Thomas C. Butler, Company H, Twenty-fifth Infantry Regiment had entered the Fort at the same time and took possession of the Spanish flag. They were ordered to give the flag to the officer of the 12th U.S. Infantry Regiment. But before releasing the flag, Butler and Jones tore a piece from the flag and kept the torn piece."

Demonstration of Black NCO Leadership

There were instances during the Spanish American War where the white commanders were absent, wounded or killed and the company leadership became the responsibility of the black NCOs. it was reported that during some battle engagements at San Juan, Cuba, units were temporarily under the commands of First Sergeants William H. Givens, Saint Foster and William Rainey.

William H. Givens assumed command of his company on July 1, 1898 when his captain was wounded in action. Givens was a veteran of 32 years having enlisted in 1866. When Lieutenant Roberts was wounded and Lieutenant Smith of the Tenth U.S. Cavalry was killed in action in Cuba, First Sergeant Saint Foster assumed command.

The soldiers of the Twenty-fourth Infantry Regiment were part of the Third Brigade commanded by Colonel Wikoff during the Spanish American War. A group of the Twenty-fourth Infantrymen were patrolling around the foot of San Juan Hill, several officers were wounded. The command of the Company F became the responsibility of First Sergeant William Rainey.

Tenth Cavalry
The San Juan Tragedy

Corporal John Walker, Troop D, 10th U.S. Cavalry, wrote the following account of his unit's part in the San Juan tragedy and the death of Lieutenant Jules G. Ord.

"Upon the 1st day of July, as the Tenth Cavalry went into battle at San Juan Hill against the Spanish Forces, Troop D Deployed to the left and joined Hawkins' brigade in the charge, the Sixth Infantry becoming excited and retreated. They stampeded the entire line, making the charge. Lieutenant Jules G. Ord, of the Sixteenth Infantry and Captain Bigelow of the Tenth Cavalry endeavored to rally the American forces and succeeded by their timely and rave assurances that by standing their ground and continuing the charge up the hill, victory was in store for them. Immediately after which Lieutenant Ord walked down the line toward the road leading to the city of Santiago. Upon seeing the gatling gun detachment selecting a more advantageous position from which to play upon the Spanish lines, and becoming greatly encouraged at this, he hastily retraced his steps down the line saying: "Men, for God's sake raise up and move forward, for our gatling guns are going to open up now." As the gatling gun opened fire upon the enemy's trenches, the Tenth Cavalry and the Sixteenth Infantry arose from their reclining position and charged forward, commanded by Captain Bigelow and Lieutenant Ord respectively.

In the charge Captain Bigelow fell pierced by four bullets from the enemy's guns. Upon falling he implored: "Men, don't stop to bother with me, just keep up the charge until you get to the top of the hill." Captain Bigelow's fall left Lieutenant Ord in command of the front forces in the charge as they ascended San Juan Hill. As we reached the Spanish trenches at the top of the Hill, Lieutenant Ord with two privates of the Sixteenth Infantry and I being the first to reach the crest, and at that time the only ones there captured four Spaniards in their entrenchments one of whom was armed with a side arm [revolver] which I took from him. Lieutenant Ord said: "Give it to me as I have lost mine, and we will proceed to this blockhouse and capture the rest of the Spanish soldiers." Taking the revolver from my hand, he and I walked toward the blockhouse. Lieutenant Ord stopped near a large tree, directing his attention to the firing which was coming from the Spaniards who had previously occupied the blockhouse fortification. Just as he tip-toed to see over the high grass, Lieutenant Ord was shot through the throat by a Spanish soldier who lay concealed in the heavy underbrush at the foot of the tree by the side of which he paused to watch the Spanish firing. As Lieutenant Ord fell upon the spot, the Spaniard jumped up and ran toward the already retreating Spanish line.

As he started to run, I shot him twice in the small of the back, killing him, one bullet entering close to the other. I was by Lieutenant Ord's side when he received the mortal wound and he fell at my feet. Without moving out of my tracks. I fired twice at the fleeing Spaniard while standing directly over Lieutenant Ord, and just before he gasped his last, he muttered: "If the rest of the Tenth Cavalry were here, we could capture this whole Spanish Command." Corporal John Walker, Troop D."

Reference: Black Defenders of America.

CHAPTER 9

BOXER REBELLION AND PHILIPPINE INSURRECTION

The Buffalo Soldier's Presence in the Chinese Boxer Rebellion, 1900

After the war between China and Japan 1894, a great dislike for Europeans and foreigners by the Chinese was increasing. There was an emergence of secret anti-foreign societies. An organization known as the "Boxers" were successful in initiating some uprisings in Northern China. They attacked hundreds of Europeans and thousands of Chinese Christians. Some foreign nations became quite concerned about the Boxer's aggressive actions. A unified foreign movement was started to occupy certain areas and provinces in China. England, France, Germany and Russia seized and occupied some areas. The United States became concerned about the Chinese government secretly assisting the Boxer Movement, American residents and their lives and property.

The English had sent a Squadron of ships to China and the United States dispatched some vessels along with Germany, Russia, France, Italy, Austria, and Japan war ships to Chinese Ports.

An International Expeditionary Force was organized including the countries of England, France, Germany, Russia, Japan and the United States. This unified forces' mission was to free the foreign citizens who were under seize in foreign legations by the Chinese during the Boxer Rebellion. The American land force under the command of Major General Chaffie arrived in Toku, China on July 28, 1900. The troops assigned to this special task force were the Fourteenth Infantry. Battery of Third U.S. Artillery, Fifth and Seventh Artillery, First and Third U.S. Cavalry. Unfortunately very few history text books have printed these missing pages.

During the Boxer Rebellion in China in 1900. We were there, people of color, part of the American land forces. There were some members of the following units who were present in china. They were the *Ninth U.S. Cavalry Regiment, Twenty-fourth Infantry Regiment and Twenty-fifth Infantry Regiment.*

Philippine Insurrection Black Cavalry Officer In the Philippines

Charles Young the third black graduate of the West Point Military Academy was assigned to the Philippines in 1901. He was promoted to Captain on February 2, 1901. Young was assigned to a subpost, Daraga on August 16, 1901. Around July 23 through August 2, 1902, he was in command of his unit. When he was transferred to San Joaquin, he assumed command of the post on June 4, 1902. He commanded the post and Troop I, Ninth Cavalry with 83 men present, a surgeon, a commissary sergeant and signal corpsman.

After conducting a reconnaissance of the post area and observing the town citizens of San Joaquin, Young wrote:

"The people of San Joaquin have never been in rebellion, are pacific laborers and poor. The president and priest are "Cabosor," leaders of the town and are gentlemen. The leaders of all the people are friends of the Americans. This town is well deserving the stationing of troops here both by reason of climatic conditions, good water and character of the people.

Captain Young enjoyed a constructive tour of duty during the Philippine Insurrection. He commanded troops at Samar, Blanca Auora, Daraga, Toboca, Rosana and San Joaquin from July 1901 to October 6, 1902. Young participated in numerous engagements against the insurgents on the Island of Samar.

Black Gallantry in the Philippine Islands

When the United States dispatched military forces during the Philippine Insurrection, there were blacks among the troops. Here again, the black soldier was on the scene, making gallant contributions that in some cases were not reported in the pages of history. A speech in the House of Representatives on Monday, June 8, 1914 by the Honorable Martin B. Madden (Illinois) relates an episode from the outstanding record of Negro troops in the Philippines. Some of the historical data for his speech was researched by a Mr. Daniel Murray, a former assistant librarian, Library of Congress for more than fifty years.

The following account depicts the bravery of American Negro soldiers under the leadership of a capable and heroic commander.

During the Philippine trouble it is related by Dr. Joseph M. Heller, late major and surgeon, United State Army, that during the campaign Captain Batchelor, a North Carolinian by birth and a hero if ever there was one, with 50 colored troopers, a brave and splendidly disciplined little band, marched and fought their way over a distance of 310 miles in one month. The route selected was over roads so difficult as to be almost impossible to travel. In fact, the route did not really deserve the name of roads, but was simply trails, through which the men plodded along, sinking at times to their knees in mud.

The expedition at the time was chasing Aguinaldo through the northern and central portions of Luzon and toward the China Sea. Dr. Heller stated that he never saw men show truer courage than those troops with Captain Batchelor. They were insufficiently clothed for the long march, and without guides in a strange region, but through chilling nights and sweltering days they forded 123 streams and crossed precipices and mountains where the daily average of ascent and desent was not less than 8,000 feet. For three weeks these troops lived on unaccustomed and insufficient foodstuffs and drove the enemy twice from strong position. They captured many of the Natives and set free more than four hundred prisoners. They finally forced the surrender of the commander of the insurrecting forces and made the people of Luzon enthusiastic advocates of American supremacy. No other single command during the Philippine trouble stood as many hardships or accomplished so much as these Negro soldiers under Captain Batchelor. Such was the report made at the time; and although General Lawton was killed, Captain Batchelor carried out his verbal orders, and died of cholera in the Philippines, thus going to his grave without any further reward or recognition for one of the bravest expeditions ever attempted by soldiers in modern times."

CHAPTER 10

PUNITIVE EXPEDITION MEXICO

Tenth Cavalry Regiment, The Truth About Carrizal

On June 21, 1916, newspapers throughout America carried the tragic news concerning a small-scale American massacre at Carrizal, Mexico. Though the only participants were three American white army cavalry officers and black troopers of Troops C and K, Tenth U.S. Cavalry, the nation was stunned and concerned. The black press and community were saddened by the incident and for some time were not made aware of the actual facts. The following account of the Carrizal incident has been summarized from official correspondence between the investigating officer of the incident, Commanding General, Punitive Expedition, U.S. Army, Mexico, and Commanding General, Southern Department, Fort Sam Houston, Texas around June to September, 1916.

A Lieutenant Colonel George O. Cress, Inspector General's Department stated the following in his report to General Pershing, Commanding General, Punitive Expedition, U.S. Army, Dublan, Mexico:

Captain Charles T. Boyd, Troop C, Tenth Cavalry, and Captain L. S. Morey, Troop K, Tenth Cavalry were each ordered by the Commanding General, Punitive Expedition, to make reconnaissance, from their respective stations, in the direction of Ahumada. There was no cooperation between these troops ordered by the Commanding General, Punitive Expedition [Cooperation by Commanders].

When Captain Boyd arrived at a the Santa Domingo ranch on June 20, 1916, he decided to assume responsibility for both troops, with Captain Morey as his subordinate. Carrizal was approximately eight miles from the Santa Domingo ranch. Ahumada was some twenty miles away. Boyd decided to pass through the town. After several conferences with Mexican officers (Lieutenant Colonel Rivas, and General Gomez), he was informed that American troops could not pass east, west or south of the area. During these conferences, the American Troops had advanced east across an open flat toward the southwest edge of Carrizal, where Mexican troops were formed. As the American troops decided to advance forward in a line of platoon column, Captain Boyd ordered Captain Morey and a Lieutenant Adair to defend their flanks. The Tenth Cavalry were facing 315 Mexican soldiers, mounted and dismounted.

The inspecting officer of the incident stated:

Captain Boyd appeared to be of the opinion that his orders required him to pass through the town of Carrizal . . Lieutenant Adair appeared to hold the same view. Captain Morey differed with Captain Boyd. . .

Finally Captain Boyd decided to go through the town according to a sworn statement by Quartermaster Sergeant Dalley Farrior, Troop C, Tenth Cavalry. Farrior said:

When we arrived near Carrizal, the captain had us load our rifles and pistols. We halted and sent a messenger in to ask permission to pass through the town. When the messenger returned, several Mexicans came with him and they halted at our point. The captain went forward and talked to them. He returned to us and said that it looked favorable but we could only go north. He said his orders were to go east and he meant to go that way. By this time, the General of the Carrizal Troops had come out and the captain went forward to talk to him. When he turned he said the general had given us permission to go through the town, but we would go through as foragers. As we formed lines of foragers the general called him back again. When he returned he said he would execute fight on foot and advance in that formation. We did this and ordered no man to fire until fired upon. As we moved forward Troop K was on the right and Troop C on the left. The captain cautioned Sergeant Winrow, who commanded the right of Troop C to keep his men on a zig zag line.

The Mexicans during this time had formed a line to our front about 200 yards away and opened fire on us. We laid down and fired back. . . Then we advanced by rushes. On the second rush I was wounded in the right arm and stood where I was. The line had been moving forward. On their third rush they reached the Mexican's first line of defense, where there were two machine guns. By this time, Captain Boyd had been shot in the hand and shoulder... The captain tried to get Troop K, which was in our rear to move up to us. He was shot and killed at that time. Lieutenant Adair had gone with his men and was out of sight. Captain Morey said to assemble Troop K on him and we would all surrender. But several men of Troop C remonstrated with Captain Morey and induced him to make towards an adobe house in our left rear, where we could make a stand. Captain Morey was very weak from loss of blood and fainted once. From here, I finally made my way to the Santa Domingo ranch.

Lieutenant Colonel Cress's conclusions of his investigation were:

That in carrying out his mission Captain Charles T. Boyd, Tenth Cavalry, did not obey the instructions given him by the Commanding General, Punitive Expedition, and that in failing to do so and in assuming command of Troop K, Tenth Cavalry, he became responsible for the encounter between the American troops and the forces of the de facto government at Carrizal, June 21, 1916.

There was not further action recommended due to the peculiar conditions at the time.

General Pershing stated in his endorsement that:

Under the circumstances, unfortunate as they were, it is not believed that any disciplinary action is indicated as advisable. There is no reliable evidence obtainable to sustain charges against any individual or group of these men for their conduct. Notwithstanding the disaster resulting from this encounter, it must be said to the credit of this little body of men that they fought well as long as their officers remained alive to lead them and for some time after...

The Mexicans sustained a loss of forty-two killed and fifty-one wounded. The following Tenth Cavalry soldiers gave their lives in the line of active combat duty at Carrizal.

Captain Charles T. Boyd, Lieutenant Henry R. Adair, Private De Witt Rucker, Troop K, Private Charles Mathews, Troop K, Sergeant Will Hines, Troop C, Lance Corporal William Roberts, Troop C, Private James E. Day, Troop K, and Private Walter Gleeton, Troop C.

Reference: Black Defenders of America 1775-1973

Buffalo Soldiers Heroes of Carrizal

The Carrizal Confrontation did arouse interest and concern in official government and diplomatic service. Seventeen American prisoners were taken by the Mexicans and later released.

After all of the survivors of the Carrizal Encounter had been accounted for, it was learned that two officers, Captain Charles T. Boyd and Lieutenant Adair had been killed and that they had been thrown into a ditch with some dead enlisted men of the troops. When the nine bodies were recovered, the bodies of the two officers and one enlisted man, Private Dewitt Rucker, Troop K, were released to relatives. The bodies of five enlisted men were sent to Arlington National cemetery for burial. It was fortunate that during this period of separate but equal standards in America, a concerned Congressman, Representative Johnson of Washington State, would introduce a resolution in the House of Representatives to provide for the military escort for the bodies of the troopers of the 10th United States Cavalry killed at Carrizal. Congressman Johnson said in the Congress that "Those in the Navy who fell at Vera Cruz were accorded similar honors and that the Carrizal victims were Negro troopers, but they were heroes the same as white soldiers."

Recently, I visited the grave sites at Arlington National Cemetery of Private James E. Day, Troop K, 10th Cavalry, Section 23, 21747; Private Walter Gleeton, Troop C, 10th Cavalry, Section 23, 21746; Sergeant Will Hines, Troop C. 10th Cavalry, Section 23, 21748; Private Charles Mathews, 10th Cavalry, Section 23, 215; Lance Corporal William Roberts, Troop C. 10th Cavalry, Section 23, 21744. While standing beside the grave sites, I thought about these five men of the 10th Cavalry and the circumstances surrounding their demise. I also wondered how many times a small flag had been placed on their grave sites through the years to remember them among the honored dead on Memorial Day. Then, I realized that to a real majority of people theirs are just one of the number of remains lying in the beautiful national cemetery.

These five men in 1994 are buried among the known, but history recalls that they were known to many on Friday, 17 July 1916, as five unidentified bodies of the 10th Cavalry regiment killed in the Mexican ambush at Carrizal. Impressive ceremonies were conducted for five men who would officially be identified by the War Department in September 1916. However, recognition, respect and military honors were rendered to these five soldiers of the 10th Cavalry. Professor John R. Hawkins, financial secretary of the African Methodist Episcopal (AME) Church, placed a wreath on each of their graves at the Arlington burial ceremony. The wreaths were placed on behalf of the "Colored Citizens Committee." The band played "Nearer My God to Thee" and "Lead Kindly Light." Chaplain Livingston Bayard of the United States Navy read the internment service. Wreaths were sent by the White House and placed upon each grave. Secretary of War Baker; Major General Hugh L. Scott, Chief of Staff, Army; Honorable L.C. Dyer, Member of Congress from Missouri; Dr. A.M. Curtis; Professor G. W. Cook, Secretary of Howard University; Judge Robert H. Terrell; John C. Dancy, former Recorder of Deeds and some Spanish War Veterans all represented some of the official civilian community present at the burial ceremony.

Ironically, it was on 10 April 1915, that Beverly Perea's dying wish was fulfilled. He became the first black known to be buried with honors in Arlington Cemetery. In 1981, the Interment Service, Administrative Office, Arlington National Cemetery, could only provide the grave site numbers, name, rank, unit and date of death and the fact that they were black. A privately published book concerning prominent and historical figures buried in Arlington, of course, did not refer to the five soldiers of Carrizal. When I passed Section 23 while on a private sightseeing tour with a narrator, I became my own narrator in order to relive the memories of not famous nor distinguished servants of America, but just five courageous men who are at rest among the known at Arlington National Cemetery.

CHAPTER 11

BIOGRAPHICAL SKETCHES

Selected Biographical Sketches
Of
"Buffalo Soldiers"

I recently made a revisit to Arlington National Cemetery and guided by the remnants of the "Separate But Equal" Period, I was able to locate some graves of the Buffalo Soldiers of Yesteryears. Many are buried in various sections that at one time were reserved for the burial of African Americans. These soldiers and in some instances their wives are really "buried Among the Known" in the national prestigious cemetery. When I located the grave site of a Medal of Honor winner from the Spanish American War, and noticed his grave marker was only 5-10 feet from the curb of a roadway. Then I understood the true meaning of my term "They Rest Among The Known" because Dennis Bell's grave is just across the street in section 33 from the entrances to President John F. Kennedy's grave site area, and also not too distance from the grave of the late Supreme Court Justice Thurgood Marshall.

I have included these brief and detailed biographical sketches to depict the factual realities that some of the soldiers had families and they were from various geographical areas of the country and above all they were laid to rest in the tradition and dignity of military service for its faithfully departed. Those brave and proud Buffalo Soldiers did answer to the "call of taps and now rest among the known at Arlington National Cemetery and many other burial places.

Henry Adams
Sergeant, Company B, Twenty-fifth Infantry Regiment

Henry Adams enlisted in the Twenty-fifth Infantry Regiment. This Buffalo Soldier died on March 19, 1912, when he responded to the "call of taps". Private Henry Adams is buried in Arlington National Cemetery.

John Hanks Alexander
Lieutenant, Ninth U.S. Cavalry Regiment

John Alexander was born on January 6, 1864 in Arkansas, the son of slave parents, James and Fannie E. Alexander. He studied at Oberlin College, Ohio but left in 1883 in his freshman year, to attend the U.S. Military Academy at West Point. Alexander graduated in 1887, number thirty-two in a class of sixty-four. His first assignment was with the Ninth Cavalry Regiment at Fort Robinson, Nebraska. In 1891, Alexander was on detached duty to inspect a unit in Raleigh, North Carolina. He was assigned to Wilberforce University, Xenia, Ohio in 1894 to serve as professor of military science and tactics. His stay at Wilberforce was brief since on March 26, 1894, on a visit to Springfield, Ohio, he died suddenly of a ruptured vessel while sitting in a barber's chair.

Professor William Sanders Scarborough, the late distinguished professor at Wilberforce told how the unfortunate death of Alexander was received.

"The President, the Secretary, Mr. Wallace Clark and I left at once for that city (Springfield, Ohio) and made arrangements for his body to be brought back to the university. The esteem in which he was held showed itself at once, the offer of the Springfield White Military guard to accompany the remains back to the University. Our Arnett Guards met the body at the city limits and took up the escort to the University where the funeral services were held".

Lt. John Hanks Alexander was the second black graduate of the West Point Military Academy and the second man of color to become a Buffalo Soldier officer.

Allen Allensworth
Lieutenant Colonel Twenty-fourth Infantry Regiment

Colonel Allen Allensworth, a Native of Louisville, Kentucky was born on April 3, 1843 of slave parents, Levi and Phyllis Allensworth. At the beginning of the Civil War, he was sold in a slave market at New Orleans for one thousand dollars to ride race horses. In the summer of 1861, he was brought back to Louisville, Kentucky by his new owner. He left Louisville in 1862 with the Union Army and obtained his freedom in 1863. On April 3, 1863, he enlisted in the U.S. Navy and advanced in rank from seaman to petty officer. He was released from the Navy on April 3, 1865. After the Civil War, Allensworth returned to Louisville and joined the Fifth Street Baptist Church. He became the janitor of the Ely Normal School in Louisville and later studied there. The Freedmen's Bureau selected him to be a teacher and school principal. He was determined to improve his education and enrolled in the Nashville Institute, later known as Roger Williams University. He pursued the normal and ministers course and then decided to leave the school. He became active within the General Association of Colored Baptists and also served as a pastor in churches at Franklin, Louisville, and Bowling Green, Kentucky.

Later, Allensworth accepted a pastorate in Cincinnati, Ohio. The Roger Williams University Subsequently conferred on him an honorary master of arts degree.

Allen Allensworth was appointed by President Grover Cleveland as a Chaplain, Twenty-fourth U.S. Infantry Regiment. He served from 1886 to 1906 and retired as a Lieutenant Colonel. In 1895, Allensworth designed a new Chaplains' insignia and forwarded it to Washington. Later, the army adopted an insignia for Chaplains that was quite similar to his design. Allensworth was the author of an outline of course of study and the rules governing fort schools of Fort Bayard, New Mexico, March 1899.

Colonel Allensworth demonstrated an interest in education. In 1891, he went to Toronto Canada to attend the National Education Association's Annual Meeting. He spoke on the "History and Progress of Military Education in the United States". He also was in attendance at 1893 World Columbian Exposition in Chicago. His last Military assignment was at Camp Reynolds on Angel Island as Post Chaplain. After his retirement from the army, he decided to live in California.

In 1908, Allensworth and William Payne, a college graduate and teacher decided to start an all black community in California. On June 30, 1908, the California Colony and Home Promoting Association was formed. John W. Palmer, a miner; William H. Peck, a minister, and Henry A. Mitchell, a real estate agent, were also involved in the Association's initial organization. The site, consisting of some twenty acres was located at Solito, approximately thirty miles from Bakersfield, California. The community was later named Allensworth. During the period 1912-1915, the community was quite prosperous. Unfortunately, Allensworth died in 1914 and the other leaders around 1920 had departed the area and the community began to decline. The deep and strong interest in education and economic survival for Afro Americans by Colonel Allensworth was not forgotten by historians and the people of California. On October 6, 1976, a state park was dedicated in the memory of the courageous minister of the gospel, educator, and military chaplain, and above all, pioneer homesteader. Today, citizens can visit and enjoy the beautiful scenery and natural surrounding in Allensworth State Park in California.

The outstanding contributions of Colonel Allen Allensworth have been honored and preserved by the naming of the state park in his memory. However, there is a stronger and more durable memorial that is needed for his words to be remembered forever.

Norman Alston
Private, Troop B, Tenth U.S. Cavalry Regiment

Norman Alston was a member of the Tenth U.S. Cavalry. Alston served in World War I. He died July 4, 1963 and is buried in the Soldiers' Home National Cemetery, Washington, D.C.

George G. Anderson
Corporal, Company K, Twenty-fifth Infantry Regiment

George Anderson was a native of Paducah, Kentucky. He enlisted in the Eighth, Illinois Volunteer Infantry Regiment and served six months in Cuba during the Spanish American War. He reenlisted in the regular army on May 18, 1899, and was assigned to Company K, Twenty-fifth Infantry Regiment. He also served for a time as headquarters clerk.

William T. Anderson
Chaplain, Tenth U.S. Cavalry Regiment

William T. Anderson was born on August 20, 1859 at Saguin, Texas and was a graduate of the Theological Department of Howard University, Class of 1886 and of the Homeopathic College, Cleveland, Ohio. Anderson was appointed Chaplain of the Tenth U.S. Cavalry Regiment on August 16, 1897, and joined the regiment at Fort Assinniboine, Montana on November 11, 1897. He was appointed post treasurer, librarian, and superintendent of the post school. At one time, he was in command of Fort Assinniboine, Montana in 1898. Later, he joined the regiment near Santiago, Cuba on July 25, 1898. Chaplain Anderson was the only known black regular army chaplain to serve in Cuba during the Spanish American War. He was able to use his medical knowledge to assist the wounded.

Sylvester Archer
Private, Troop C, Tenth U.S. Cavalry Regiment

Sylvester Archer was born in 1853 in Broome County, New York. He enlisted on April 8, 1863. The official records listed his description as five feet, six inches, brown complexion, brown eyes and black hair. His civilian occupation was a farmer. Archer was married twice.

Archer married Sarah J. Patra on April 1, 1862. She was the daughter of Hannah Beebe. Sylvester and Sarah were married by Rev. George Bosley who was the pastor of the African Methodist Episcopal Zion Church located on Whitney Street, Binghamton, New York. The church was located across the street from her mother's house. Rev. Bosley later lived at the rear, 1453 Rhode Island Avenue, Washington, D.C. The persons present at the wedding ceremony were Henry C. Williams, Catherine Davis, Leak Davis, Charles Jackson and John Holland. Sarah stated in a sworn affidavit in support of her widow's pension that Archer had a furlough for thirty days while in the army and did not return again until he was discharged in 1865 and then left in 1867 to enlist in the regular United States army. Sylvester Archer enlisted at Binghamton, New York on August 19, 1867 and was assigned to the Tenth U.S. Cavalry, Troop C. Sarah said her husband wrote letters frequently while in the service and was out West fighting the Indians. She never heard from him again until after his discharge from the Tenth U.S. Cavalry. She read of his death in a newspaper. He was discharged on August 17, 1872 at Camp Supply, Indian Territory.

Archer met Abbie Green in Strong City, Kansas and married her. Abbie was born free in Charlotte County, Virginia. She stated in a sworn affidavit that around 1862, her mother brought her to Gallia County, Ohio. In 1862, she married Jefferson Green and they were the parents of five children. The oldest child was John Henry Green. Abbie said that when she left her husband while living on a farm (homestead claim) in Strong City, Kansas that she never again heard from any of her children. She met Archer and they were married on April 18, in Topeka, Kansas. At the time Sylvester and Abbie married, neither were divorced from their legal spouses. Abbie had a sister, Mary Henry, who lived in Springfield Township, Callia

County, Ohio and was present at her wedding to Jefferson. Sylvester and Abbie were the parents of Barteen, born September 19, 1885; Gertrude, born December 7, 1887; Zephyr, born March 26, 1890; Martha, born June 30, 1892 and Henrietta, born May 20, 1894. Abbie Green did not know of Archer's previous marriage until his death. Sylvester Archer suffered with disabilities in later years of his life. They were palpitation of the heart, pleurisy, rheumatism, disease of the eye, and scurvy. He died on April 9, 1902 at Meriden, Kansas at the age of 59 years.

Sarah J. Archer had filed for a widow's pension under the Act of June 27, 1890. She was living at 26 Wilber Street, Binghamton, New York and was 62 years old when she was awarded a widow's pension from the date of Sylvester's death. A recommendation of the Adjutant General, Binghamton, New York, headquarters Union Veterans Order of the Union Battle Men Army of New York and New Jersey, said *To whom it may concern. I am glad to say that I have known Mrs. Sarah Archer at least twenty years and that she bears a good moral character. I was superintendent of the Colored People's Sunday School for fourteen years which gave me the best opportunity of knowing her.* Archer's legal wife, Sarah, died on May 8, 1911 at 55 Sherman Place, Binghamton, New York. In January, 1921, Zephyr Hysten wrote a letter to the Pension Bureau, Washington, D.C. and requested a pension for her mother, Abbie Green (alias Ida Jackson Archer), because her mother was never legally married to Sylvester Archer, it was believed that her request for a widow's pension was denied.

Richard B. Ashe
Private, Tenth U.S. Cavalry Regiment

Richard B. Ashe was from Washington, D.C. He was born on March 20, 1857. He enlisted in the Tenth U.S. Cavalry Regiment. Ashe was married to Mary Ashe who was born on November 29, 1869. Richard B. Ashe died on February 13, 1942. His wife, Mary died on May 20, 1942. Richard and Mary Ashe are buried in Arlington National Cemetery.

Lemuel Ashport
Corporal, Twenty-fifth U.S. Infantry Regiment

Lemuel Ashport was born on March 22, 1846 in Taunton, Massachusetts. He was the son of Noah Ashport born in Brockton, Massachusetts and Esther Wood Ashport born in Marshfield, Massachusetts. Lemuel enlisted on December 16, 1863 at West Bridgewater, Massachusetts. The official record listed his description as five feet, ten inches, 165 pounds, brown eyes, black-gray hair and brown complexion. His civilian occupation was a shoemaker. He signed his name.

Ashport married Elizabeth Pierce, the daughter of Albert A. and Sarah Pierce. Elizabeth had a brother, Charles E. Pierce of Bridgewater, and a sister, Rowena A. Sanders Lemuel and Elizabeth were the parents of Lillian E., born November 1, 1882, Ethel H., born September 16, 1886 and Pearl E., born January 14, 1888. Elizabeth and Lemuel were married by Rev. J.E. Hawkins in November, 1881.

Ashport was discharged on August 20, 1865. He enlisted on August 15, 1867 in Company I, 39th U.S. Infantry and was transferred to the 25th U.S. Infantry by consolidation of April 20, 1869. He was appointed corporal. Lemuel Ashport died at the age of 58 years while living at 22 French Court, Brockton, Massachusetts. The primary cause of death was a cerebral hemorrhage and the contributing cause was a fall downstairs. He died on February 29, 1905 and was buried at the Corviett (Conerett) Cemetery, Brockton, Massachusetts.

Edward L. Baker Jr.
Sergeant Major, (later commissioned Second Lieutenant),
Tenth U.S. Cavalry Regiment.

Edward Baker, a black man, was born in Laramie County, Wyoming. He was awarded the Medal of Honor on July 3, 1902, for gallant action in Cuba. The citation read: *"On July 1, 1898, at Santiago, Cuba, he left cover and under fire rescued a wounded comrade from drowning."*

Dennis Bell
Private, Troop H, Tenth U.S. Cavalry Regiment

Dennis Bell, a black man, was born in Washington, D.C., where he entered the army. He was awarded a Medal of Honor for heroic action in Cuba, the citation reading as follows: *"Bell exhibited a heroic act when he voluntarily went ashore at Tayabacca, Cuba, 30 June 1898, in the face of the enemy and aided in the rescue of his wounded comrades; this after several previous attempts had been frustrated."* Bell is buried in Arlington National Cemetery across from the John F. Kennedy gravesite.

George M. Berry Jr.
Private, Troop E., Tenth U.S Cavalry Regiment

George M. Berry Jr. was a member of the Tenth Cavalry Regiment. He was married to Henrietta Berry on April 5, 1874. She died on November 7, 1958. George M. Berry Jr. died on March 5, 1927. George and Henrietta Berry are buried in Arlington National Cemetery.

George Berry
Sergeant, Troop G, Tenth U.S. Cavalry Regiment

George Berry served with his unit in Cuba, and was instrumental in the memorable occupation of San Juan Hill. He displayed courage and initiative when he carried the colors of the Third and Tenth up San Juan Hill. It was reported that he constantly shouted to his men, *"Dress on the colors, boys. Dress on the colors"* Berry retired from the army after thirty years of service.

Gaines Billingslea
Private, Tenth U.S. Cavalry Regiment

Gaines Billingslea was a member of the Tenth U.S. Cavalry Regiment. He was from the state of Georgia. He died on July 26, 1949 and is buried in Arlington National Cemetery.

Horace Wayman Bivins
Sergeant, Tenth U.S. Cavalry Regiment

Horace Bivins was born in Pungoteague, Accomack County, Virginia, on May 8, 1862. He was raised on a farm and at the age of fifteen was given the responsibility of running an eight-horse farm one mile from Keller Station, Virginia. On June 13, 1885, Bivins enrolled at Hampton School where he received his first military training. On November 7, 1887, he enlisted in the army at Washington, D.C. He was sent to Jefferson Barracks, Missouri, and after training, on June 19, 1888, was assigned to Troop E, Tenth U.S. Cavalry, at Fort Grant, Arizona Territory. Bivins served with this troop on various missions around San Carlos, in the Arizona Territory and when the unit arrived at Fort Apache, Bivins was detailed as a clerk in the regimental adjutant's office. He held this position from November 10, until June 15, 1890. On April 14, 1898 his regiment was ordered to Chickamauga to prepare for participation in the Spanish American War Cuban campaigners. Sergeant Bivins was a successful sharpshooter and at San Carlos participated in his first rifle competition. He gained second place in a troop of sixty men. In 1889, he was made a sharpshooter and represented the troop in 1892, 1893 and 1894, winning eight medals and badges in departmental competition. In 1894, Bivins won three gold medals representing the Department of the Dakotas at the army competition at Ft. Sheridan.

George M. Blackburn
Sergeant, Ninth U.S. Cavalry Regiment

George M. Blackburn enlisted in the Ninth U.S. Cavalry. He was born March 20, 1867. Blackburn died on December 19, 1925. He is buried in Arlington National Cemetery and responded to the "call of taps".

Joseph A. Blackburn
Troop L, Tenth U.S. Cavalry Regiment

Joseph A. Blackburn was born in 1851. He enlisted on September 10, 1867 in the Tenth U.S. Cavalry. He was stationed in Jacusbaro, Texas area, Fort Concho (San Angelo, Texas) and Fort Sill, Indian Territory. Blackburn served on the frontier during the Indian campaign.

Joseph J. Blakeney
Sergeant, Twenty-fourth U.S. Infantry Regiment

Joseph J. Blakeney was a member of the Twenty-fourth U.S. Infantry. Blakeney was from the state of Arkansas. He served in World War I. Joseph J. Blakeney died October 24, 1971. He is buried in the Soldiers Home National Cemetery, Washington, D.C.

Jackson Bogle
Private, Company D, Twenty-fourth Infantry Regiment

Jackson Bogle was a member of the Twenty-fourth Infantry Regiment. Bogle served in the Spanish American War. He died on September 11, 1915. Bogle is buried in Arlington National Cemetery.

John H. Bowman
Private, Company B, Tenth U.S. Cavalry Regiment

John H. Bowman enlisted in the Tenth U.S. Cavalry and served in the Spanish American War. This Buffalo Soldier died on March 5, 1904, when he responded to the "call of taps". Private John H. Bowman is buried in Arlington National Cemetery.

Thomas Boyne
Sergeant Company C, Ninth U.S. Cavalry Regiment

Thomas Boyne was from Prince Georges County, Maryland. During the Civil War, he was a member of a light artillery regiment and served in Texas. He was discharged in 1866 at Brownsville, Texas. Boyne enlisted in the Twenty-fifth Infantry Regiment in 1867. Later, he joined the Ninth Cavalry Regiment and retired in 1889.

Boyne's outstanding gallantry in action occurred while he was leading a detachment of men through the Mimbres Mountains, New Mexico on May 29, 1879 and encountered some Mesecalero and Chiricahua Apaches Indians led by Chief Victorio. When Boyne's commanding officer, Captain Charles D. Beyer from Fort Bayard was assisting a wounded man and was surprised by some Indian warriors, Boyne was in charge of a detachment that came to Beyer's rescue. Sergeant Boyne flanked the charging Indians and firing with precision, he eventually drove the Indian warriors away. Later, on September 29, 1879 at Rio Cuchillo Negro, New Mexico near Ojo Caliente the Ninth Cavalry confronted the Indian warriors and fought a running battle for two days. Boyne, though wounded, killed one of three warriors and single handedly captured a considerable number of the sixty horses and mules that were taken. His gallantry in action during the encounter earned him the Medal of Honor. He was awarded the medal on January 6, 1882.

William P. Brooks
Corporal, Troop G, Ninth U.S. Cavalry Regiment

William P. Brooks was a member of the Ninth U.S. Cavalry. Brooks was born October 21, 1878. He served during the Spanish American War. Brooks died August 3, 1969. He is buried in the Soldiers' Home National Cemetery, Washington, D.C.

Arthur Brown
Surgeon Tenth US Cavalry Regiment

Arthur Brown was born in Raleigh, North Carolina. He completed his secondary education in Raleigh and enrolled in Lincoln University, Pennsylvania, where he graduated in 1888. Brown received a degree Doctor of Medicine from the University of Michigan at Ann Arbor in 1891. He began his practice in Birmingham, Alabama. He opened a drug store and his medical practice. When the Spanish-American War commenced, Brown organized a company of volunteers. The company was not activated. Brown decided to write a letter to Doctor A.M. Curtis, a distinguished physician in Washington, D.C. Arthur Brown received a letter from the Surgeon Chief Sternberg and received an appointment as a surgeon and ordered to report to an immune regiment in Cuba. He was assigned to the Tenth Cavalry Regiment. He was known to be one of the most active surgeons on the battlefield in Cuba. Brown was the only physician of color assigned to Cuba during the Spanish American War. It has been stated that Dr. Arthur Brown was in "command of the entire regiment from August 12, to October 8, 1898." That possibly could have occurred in the absence of Lieutenant Colonel T. A. Baldwin who commanded the Tenth Cavalry during the period May 10, 1898 to November 21, 1898. The author of *Under Fire with the Tenth US Cavalry* stated that during the command of Brown, perfect peace and harmony reigned in the regiment. This outstanding Buffalo Soldier whose medical expertise and skills were used in treating patients with yellow fever and combat wounds believed in upward mobility.

Benjamin Brown
Sergeant, Company C, Twenty-fourth U.S. Infantry Regiment

Benjamin Brown was born in Spotsylvania County, Virginia. On May 11, 1889, Brown was on escort duty with a detachment escorting a U.S. army pay master, Major Joseph W. Wham. The detail was attacked by a group of robbers between Cedar Springs and Fort Thomas, Arizona. The white civilian robbers had placed a large boulder onto the road to stop the payroll convoy. When the escort detail challenged the robbers who eventually escaped with $28,345.10, Sergeant Brown although shot in the abdomen, did not leave the scene until again wounded in the same area. Sergeant Brown was awarded the Medal of Honor for his heroic action on February 19, 1890. Brown suffered a stroke while stationed at Fort Assinniboine, Montana in 1904 and his disability led to his retirement. In 1905, Benjamin Brown was living in the U.S. Soldiers Home, Washington, DC. An outstanding Buffalo Soldier and a hero died at the Soldiers Home in 1910.

David Brown
Trooper, Ninth U.S. Cavalry Regiment

David Brown enlisted in the Ninth U.S. Cavalry Regiment. He was the son of David P. and Charlotte A. Brown. He lived in Philadelphia, Pennsylvania prior to his family moving to Worcester, Massachusetts in 1850. David had two sisters, and a brother. Hattie who was married to a Spencer Winn, they were the parents of Bob and Hattie Winn. Her other sister was Henrietta. His brother was John S. Brown who served in the Civil War as a member of the Fifty-Fourth Massachusetts Regiment. David Brown was married to a white woman named Hattie. Trooper David Brown, an early "Real Buffalo Soldier" died in Cambridge, Massachusetts in 1902.

George Brown
Private, Ninth U.S. Cavalry

George Brown enlisted in the 9th Cavalry Regiment. He was from Massachusetts. Brown was born on March 26, 1879. He died at the age of 71. Brown is buried at Arlington National Cemetery.

Plummer Brown
Corporal, Twenty-fifth U.S. Infantry Regiment

Plummer Brown was born in 1854. He was the son of Willis and Charlotte Brown. His parents were former slaves from Jackson, Tennessee. They migrated with five children to Oberlin, Ohio in the early 1860's. Willis Brown died in January, 1863 at 33 years of age, leaving his wife a widow with six children. When Plummer enlisted in the Twenty-fifth Infantry Regiment in 1875, he was able to provide some support for his mother.

Plummer Brown was stationed at Fort Concho, Texas and Fort Meade, Dakota Territory in the 1880's. He became ill with consumption or tuberculosis and was discharged on November 4, 1884. While returning home to Oberlin, Ohio by stage coach his illness worsened. Brown died shortly after arriving at his mother's residence in Oberlin, Ohio.

Alexander J. Bruton
Sergeant, Company G, Twenty-fourth Infantry Regiment

Alexander J. Bruton enlisted in the Twenty-fourth Infantry. Bruton was from the state of Alabama. He served in World War I. Alexander J. Bruton died November 25, 1972 and is buried in the Soldiers' Home National Cemetery, Washington, D.C.

Benjamin Buckner
Sergeant, Tenth U.S. Cavalry Regiment

Benjamin Buckner was a member of the Tenth U.S. Cavalry. Buckner died August 27, 1922. He is buried in the Soldiers' Home National Cemetery, Washington, D.C.

William H. Burn
Private, Ninth U.S. Cavalry Regiment

William H. Burn was a native of Kentucky. He was born on January 15, 1929. Burn enlisted in the Ninth U.S. Cavalry Regiment. He was married to Virginia Burn who was born on September 3, 1877. William and Virginia Burn are buried in Arlington National Cemetery.

John Burton
Private, Tenth U.S. Cavalry Regiment

John Burton enlisted in the Tenth U.S. Cavalry Regiment, Company L on September 3, 1867. When he was discharged, he reenlisted on September 3, 1872 and served during the Civil War in Company B,
Second U.S. Cavalry Regiment, United States Colored Troops (USCT). Burton was stationed at Fort Sill and Fort Richardson Burton was born in 1844 and in 1907 was living in San Antonio, Texas. Burton was a Buffalo Soldier who served during the Indian Campaign.

James W. Bush
Trooper, Ninth US Cavalry Regiment

James W. Bush was born on August 28, 1843, in Lexington, Kentucky in 1842. He enlisted on May 12, 1863 for three years at Readville, Massachusetts by R.P. Hallowell and Captain Collins. The official records listed his description as 21 years, five feet five and one-half inches, dark complexion, brown eyes and black hair. His civilian occupation was a student in Xenia, Ohio (possibly Wilberforce University). He was married.

Bush was injured in the leg while tearing up railroad tracks at Wateree Junction, (then the branch of the Charleston and Savannah railroads), South Carolina. A former comrade stated the following in a sworn statement *I, Joseph A. Palmer, former Sergeant, Company K, Fifty-Fourth Massachusetts, Regiment was present when we were in a raid through South Carolina from Georgetown to Camden, South Carolina to dispense some enemy weapons and destroy the railroad and rolling stack, part of our company was ordered to destroy a train of cars loaded with arms, ammunition and a army supplies. It was 12 miles from camp, I remember that the company returned without the First Sergeant Bush and he was absent from the company several days. His duties were performed by the next sergeant in command. Bush was assisted back to camp.*

After his discharge from the army in 1865 Bush returned to Lexington, Kentucky. He enlisted on December 20, 1866 at Lexington and was assigned to Troop H. Ninth U. S. Cavalry. In 1870 he transferred to Troop I, and in 1872 to Troop M. He served in the Ninth U.S. Cavalry during the period 1866-1882, a total of 16 years plus 2 years during the Civil War, a total of 18 years active military service. He served at the following locations with the cavalry: Fort Stockton, Texas; Guadaloupe, New Mexico; Fort Quitman, Texas; Fort Davis, Texas; Fort Concho, Texas; Fort Stanton, New Mexico; and, Fort Union, New Mexico. He also performed field duty at Roswell, New Mexico.

Bush received another injury to his leg while serving in the cavalry near Fort Stanton, New Mexico on August 28, 1878. He was injured in the right leg while on a charge in the pursuit of some desperados or bandits. The injury eventually resulted in a tumor on the right leg. He made an application for a disability pension in 1883 while living at 663 South 20th Street, Concordia, Kansas for injuries received while in the Civil War and service with the 9th US. Cavalry. He later wrote the commissioner of Pensions, Washington, DC on September 11, 1905 and said *I have faithfully served the cause of the government and stood by this government for 17 years, 8 months and 13 days. I had to leave the service with four honorable discharges, one from Company K, Fifty-Fourth Massachusetts Regiment and one each for Troops I, M and H, US Ninth Cavalry. Bush retired from the Cavalry on August 22, 1882. He also*
explained in his letter how he received the leg tumor. He wrote *The tumor was caused while on detachment service under the command of First Lt. M. F. Goodwin, Ninth US Cavalry. I was told to prepare my horse at once for field duty which I did, I left Fort Stanton, New Mexico at 12 'clock that night for Eagle Creek, New Mexico after a band of desperados, namely Bill Kid, Dr Spurlock, Charles Bodney, McSwain, Dolan, Murphy and others. During the force and inclement night and while on a long scout under the command of the 15th U.S. Infantry. We were pursuing the famous Indian Chief Victoria and Nama in the Republic of Mexico.*

Bush was previously married during slavery. He was married at the age of 20 years. He did not marry again until 1883. He married Alice Curtis Bush on July 6, 1883 at Concordia, Kansas. Alice had a brother George Curtis of Lincoln Nebraska. James Bush died on April 12, 1918, at Lincoln, Nebraska

John P. Campbell
Staff Sergeant, Ninth U.S. Cavalry

John P. Campbell enlisted in the Ninth U.S. Cavalry in 1912 and was assigned to Ft. Huachuca. Campbell served on border patrols and also served in World War I and World II. He retired in Phoenix, Arizona.

John C. Cantey
Corporal, Company G, Twenty-fourth Infantry Regiment

John C. Cantey enlisted in the Twenty-fourth Infantry Regiment. He was born on December 15, 1879. Cantey is buried in Arlington National Cemetery.

Beverly Carter
Private, Company H, Tenth U.S. Cavalry Regiment

Beverly Carter was a member of the Tenth U.S. Cavalry Regiment. Carter died on August 24, 1894, when he responded to the "call of taps." Private Beverly Carter is buried in Arlington National Cemetery.

Isaiah Carter
Private, Troop A, Ninth U.S. Cavalry Regiment

Isaiah Carter enlisted in the Ninth U.S. Cavalry Regiment. Carter served during World War I. He died July 4, 1958. Carter is buried in the Soldiers Home National Cemetery, Washington, D.C.

Louis Augustus Carter
Chaplain, Ninth and Tenth U.S. Cavalry Regiments, Twenty-fourth and Twenty-fifth Infantry Regiments

Louis A. Carter was born in Auburn, Alabama on February 20, 1876. His education consisted of attending Tuskegee Institute, Tuskegee Alabama, Selma University, Selma Alabama and he received his bachelor of Divinity degree from Virginia Union University, School of Technology, Richmond, Virginia. In 1907, The Guadalupe College of Texas awarded him a Doctor of Divinity (DD) degree. Carter was married to Mary B. Moss in 1909.

Carter pastored churches in Dawkins, Alabama, Ashland, Virginia, and Knoxville, Tennessee prior to accepting an appointment as a regular army chaplain. Carter gave outstanding performances as a chaplain for thirty years. He was the only known chaplain to serve with all four black army regiments.

Chaplain Carter served at the following military camps or forts: Sackett Harbor, Madison Barracks, New York, Fort Benning, Georgia, Camp Stephen D. Little, Nogales, Arizona, Camp Stotsenburg, Philippine Islands, Fort Ethan Allen, Vermont and Fort Leavenworth, Kansas.

Carter demonstrated a great interest in the welfare, spiritual and upward mobility of the personal lives of the men assigned to his respective commands. Chaplain Carter would visit the hospitals, guard homes, recreation and educational facilities in order to provide guidance, counsel and gave assistance to the men assigned at posts where he was present. Chaplain Carter supported the National Association For the Advancement of Colored People (NAACP) and W.E.B. Dubois, editor of the Crisis Magazine. Carter encouraged the troops to read the Crisis. Sometimes he would disagree with W.E.B. DuBois.

Chaplain Louis Augustus Carter was the recipient of the Mexican Border Service Medal, World War I Victoria Medal and Expert Badge with Pistol bar. Carter was the first regular army black chaplain to be promoted to the rank of "colonel". Western University of Kansas awarded Chaplain Carter a "Doctor of Divinity on April 27, 1936.

A concerned spiritual leader, military chaplain and indeed a "Buffalo Soldier" Louis Augustus Carter responded to the sacred strains of "Taps" when he was buried in the Fort Huachuca Cemetery. Chaplain Carter died in a Veterans Hospital on July 16, 1941, in Tuscon, Arizona.

Martin Cashwell
Sergeant, Tenth U.S. Cavalry Regiment

Martin Cashwell was a native of Virginia. He enlisted in the Tenth U.S. Cavalry. Cashwell died in 1940 and is buried in Arlington National Cemetery.

John Causby
Private, Twenty-fifth Infantry Regiment

John Causby was born on June 9, 1880. He enlisted in the Twenty-fifth Infantry Regiment. Causby served in Cuba during the Spanish American War. John Causby was married to Minnie Causby who was born on September 27, 1897. John Causby died at the age of 79 years on April 5, 1959. His wife died at the age of 50 years on July 27, 1947. John and Minnie Causby are buried in Arlington National Cemetery.

Frank Coalman
First Sergeant, Twenty-fifth Infantry Regiment

Frank Coalman enlisted in the Twenty-fifth Infantry Regiment. Coalman served in Cuba during the Spanish American War. Frank Coalman is buried in Arlington National Cemetery where he responded to the "call of taps".

Benjamin Cochran
Private, Ninth U.S. Cavalry Regiment

Benjamin Cochran was born in 1845. He was living in San Antonio, Texas in 1907. Cochran enlisted in the Ninth U.S. Cavalry Regiment in 1866 and was assigned to Company B. He was discharged in 1871. Cochran was a Buffalo Soldier who served during the Indian Campaign.

George Conrad Jr.
Private Tenth US Cavalry

George Conrad Jr. was born February 23, 1860, a slave in Connersville, Harrison County, Kentucky. His master was Joe Conrad who served in the Confederate Army during the Civil War. After he was freed, George Conrad enlisted at Fort Riley Kansas in the Ninth US Cavalry on October 1, 1883. He was assigned to Troop G and was stationed at Fort Sill, Oklahoma at one time. In the 1930's Conrad was living in Oklahoma City, Oklahoma.

William T. Conray
Corporal, Troop H, Ninth U.S. Cavalry Regiment

William T. Conray was a member of the Ninth Cavalry Regiment. He served in Cuba during the Spanish American War. He was born on December 18, 1864 and died on October 18, 1964. Conray was married to Martinia H. Conray who died on August 11, 1944. William and Martinia Conray are buried in Arlington National Cemetery.

Eli Cook
Private, Ninth U.S. Cavalry

Eli Cook was a member of the Ninth Cavalry Regiment. He was born on August 15, 1867. Cook died at the age of 85, on February 7, 1952. "He Rests Among the Known" at Arlington National Cemetery.

William H. Cook
Private, Troop C, Tenth U.S. Cavalry Regiment

William H. Cook enlisted in the Tenth U.S. Cavalry Regiment. He was married to Minnie Cook who was born on October 28, 1885. William Cook died on December 28, 1940. His wife Minnie died at the age of 91 years on January 19, 1976. William and Minnie Cook are buried in Arlington National Cemetery.

Floyd Henry Crumbly
Sergeant Major, Tenth US Cavalry Regiment

Floyd Henry Crumbly was a native of Rome, Georgia. He was married to Rebecca Tate Crumbly. Crumbly enlisted in the Tenth Cavalry Regiment on November 16, 1876. He demonstrated a desire for advancement in the military service. Crumbly served as a company clerk, quartermaster sergeant and post sergeant major at Fort Stockton. He served during the Indian Campaigns in the West to include the "Northern Cherokee Campaign of 1878, Comanche Campaign, and the Victoria War 1880".

Dr. Charles Johnson, Jr., Chairman Department of History Morgan State University, Baltimore Maryland states in his book, *African American Soldiers in the National Guard* that Floyd Henry Crumbly after his discharge in 1888 from the Tenth Cavalry Regiment, enlisted in the 2nd Battalion, Atlanta Georgia on January 21, 1891. During the Spanish American War, Crumbly was given a Lieutenant's Commission in the Tenth Immune Regiment on July 26, 1887. According to Johnson, Crumbly received a recommendation for service overseas, from Booker T. Washington of Tuskegee Institute and other distinguished leaders and officials. In September 1899, Floyd Henry Crumbly was commissioned a captain in the Forty-ninth Volunteer Regiment and served in Paranague Province, Philippine Islands as a province commander and judge of a municipal court. After his discharge at Presdio, California in 1901, Crumbly opened an employment and real estate office in Los Angeles".

Floyd Henry Crumbly was a "Real Buffalo Soldier" who believed in the pursuit of excellence in his military career. His promotions through the years from company clerk to captain is indicative that African Americans of the former Buffalo Regiments of the West did have some members who pursued success beyond others expectations.

Charles Cunningham
Trooper, Ninth US Calavry Regiment

Charles Cunningham was born in Franklin County, Pennsylvania in 1844. He was the son of Mary Ann Brady. Charles had a sister named Jennie V. Cunningham. He enlisted in the Fifty-Fourth Massachusetts Regiment on April 8, 1863 and was discharged August 20, 1865. The official records listed his description as *five feet, eight inches, black complexion, black eyes and black hair*.

Cunningham had several marriages. He married in 1868, Frances Smith in Middleton Dauphin, Pennsylvania by E. Shafer. They were the parents of Sarah Williams born in 1870 and Charlotte Finkbone, born around 1872. He had married a Mary Ray who was deceased. On April 17, 1908, a Charlotte Sarah Cunningham signed a sworn deposition and said, *She married under the name of Charlotte Sarah Coleman while living in Kansas City, Kansas, Wyandotte County to Charles Cunningham.* The Probate Judge was K.P. Snyder and the records were certified by Probate Judge Winfield Truman on November 11, 1901.

Charles Cunningham reenlisted in the United States Army on December 16, 1872 and was assigned to Troop K, United States Ninth Cavalry. He served with the Ninth U.S. Cavalry in the following areas: Fort Lewis, Colorado; Fort Henry, New Mexico; Fort Stanton, New Mexico; Fort Bayard, New Mexico; Rock Creek, Colorado; and Fort Wingate, New Mexico.

Cunningham was wounded on January 1, 1880 by a Mexican assassin who attacked him while asleep in the line of duty as a mounted courier between the Rio Peidea and Animas City, Colorado. He was treated at the military hospital for his injuries. The wound was *an incision of the face extending from the inside corner of the left eyelid downward and inwardly through the nose and cheek inflicting a deep wound by an axe in the hands of a Mexican at Pedro's Ranch.*

Charles Cunningham was discharged because of his disability on June 6, 1888. He died on August 8, 1906 from Brights Disease.

Charles Daniel
Private, Troop K, Ninth U.S. Cavalry Regiment

Charles Daniel enlisted in the Ninth U.S. Cavalry Regiment. He served in the Spanish American War. Daniel is buried in Arlington National Cemetery.

Benjamin Oliver Davis Sr.
Brigadier General US Army, Tenth US Cavalry Regiment

Benjamin O. Davis Sr. was born on July 1, 1877, in Washington, DC. He was the son of Louis P.H. and Henrietta Steward Davis. He married Elnora Dickerson of Washington in 1902 (deceased 1916) and later married Sadie E. Overton of Mississippi, in 1919. He had three children, Olive Elnora (Mrs. George W. Streator) Lieutenant General Benjamin O. Davis Jr. (Ret) and Elnora Dickerson (Mrs. James A. McLendor). General Davis attended the Washington public schools and Howard University. He was commissioned a first lieutenant in the Eight U.S. Volunteer Infantry and served from 1898-1899. He was mustered out, reenlisted as a private and became a squadron sergeant in the Ninth Cavalry, 1899-1901. When Colonel Charles Young was stationed at Fort Duchesne, Utah (then a lieutenant) where the Third Squadron of the Ninth was stationed, he became very interested in Sergeant Major Davis' career. Young encouraged Davis to study and take the examination for lieutenancy. Even the white officers of the regiment encouraged Davis to take the examination. They offered him the necessary instruction. Colonel Young tutored Davis in the discipline of history, geography, international law, mathematics, surveying and drill regulations. On February 2, 1901, he was commissioned a second lieutenant in the cavalry, regular army. He was promoted periodically during the period 1901-1930 and attained the rank of general in 1940. He was commanding officer of the Three Hundred and Sixty-Ninth Infantry New York National Guard, 1938-1940. Davis served with the Tenth Cavalry in Samar and Panay, Philippine Islands during the insurrections of 1901-1902. General Davis was a military attache at the American Legation, Monrovia, Liberia, from 1911 to 1912. He was in the Mexican Border Patrol from 1915-1917. Davis served as professor of Military Science at Wilberforce University and Tuskegee, Alabama. General Davis was the commanding general of the Fourth Cavalry Brigade from 1941-1944.

During World War II he was special advisor and coordinator to the theatre commander, European Theatre of Operations, 1944-1945. In 1945, he was appointed an assistant to the inspector general and later a special assistant to the secretary of the army. He was awarded a scroll by President

Roosevelt for his fifty years service in the US Army. General Davis retired from the Army in 1948. He was appointed by the president as a special representative (with rank of Ambassador Extraordinary Plenipotentiary) to the First Centennial of Independence, Republic of Liberia.

General Davis' many decorations and awards have included Spanish American War, Philippine Instruction, Mexican Border, World War I and World War II service medals: the distinguished service cross, the Bronze Star Medal and the French Croix de Guerre with Palm. He was the first African American general in the United States military organization since the reconstruction period, and was actually the first black general in the active regular army. During reconstruction, the black generals were in the state militia or national guard. Brigadier General Benjamin Oliver Davis Sr. an outstanding General Officer and former Buffalo soldier died on November 26, 1970 at the Great Lakes Naval Hospital.

John Denny
Sergeant Troop B Ninth US Cavalry Regiment

John Denny was born at Big Flats, New York. He enlisted in the Ninth US Cavalry at the age of 24 years in 1867, at Elmira New York. On September 18, 1879, at Las Animas Canyon, New Mexico, Denny showed unusual heroism in carrying a wounded comrade, Pvt. Freeland, under fire to a place of safety, John Denny was awarded the Medal of Honor on November 27, 1894, for this gallant feat. Denny retired from military service in 1897. He lived in Baltimore, Maryland and later at US Soldiers Home Washington, DC where he died in 1901.

Charles S. Dorsey
Private, Company F, Twenty-fifth Infantry Regiment

Charles S. Dorsey was born on September 15, 1876. He was a member of the Twenty-fifth Infantry Regiment. Dorsey died on June 5, 1926 at the age of 50 years. He is buried in Arlington National Cemetery where he "Rest Among The Known".

Edward Dorsey
Private, Troop K, Tenth U.S. Cavalry Regiment

Edward Dorsey was born in 1871. He served during the Spanish American War. Dorsey died in 1929 and is buried in Arlington National Cemetery.

Stephen H. DuBoise
Private, Tenth U.S. Cavalry Regiment

Stephen H. DuBoise enlisted in the Tenth U.S. Cavalry. DuBoise was a cook. He died September 19, 1927. DuBoise is buried in the Soldiers' Home National Cemetery, Washington, D.C.

Elijah Duffin
Private, Twenty-fifth Infantry Regiment

Elijah Duffin was a member of the Twenty-fifth Infantry Regiment. He was married to Emma Duffin, who was born in 1885. She died at the age of 80. Elijah Duffin died on January 22, 1952. He and his wife are buried at Arlington National Cemetery.

Robert L. Duvall
First Sergeant, Company D, Twenty-fourth Infantry Regiment

Robert L. Duvall was a member of the Twenty-fourth Infantry Regiment He served in Cuba during the Spanish American War. Duvall was married to Edna Duvall who was born on August 24, 1884. Robert L. Duvall died on February 10, 1930. His wife Edna died at the age of 86 years on January 16, 1970. Robert and Edna Duvall are buried in Arlington National Cemetery.

Pompey Factor
Private Detachment Seminole Indian Scouts US Army

Pompey Factor was born in 1849 in Arkansas. He enlisted in the Army at the age of 21 years at Fort Duncan, Texas. He was later discharged but reenlisted during the period 1871-1879. Official records show that Factor served with forces mobilized for operations against Indian warriors in Texas in 1871, 1872, and 1875. He was with a detachment of scouts in an engagement with Comanche Indians on April 25, 1875, at the mouth of the Pecos River, in Texas. During the engagement, Factor assisted his comrades Isaac Payne and John Ward in rescuing a Lieutenant Bullis from a group of advancing Indian warriors. Factor was awarded a Medal of Honor for his part in this rescue. His description on official record was: height; five feet, six inches, complexion; black, hair black and eyes: black.

Pompey Factor possibly affected by the killing of a fellow Seminole Negro Indian scouts, Adam Paine by a white sheriff, decided to desert and go to Mexico. While in Mexico, Factor fought the Indians for the Mexican government. He surrendered to military authorities on May 25, 1879. The Army was lenient toward him and restored him to duty. Factor last enlisted in 1880 and returned to Musquiz Mexico and Brackettville, Texas where he worked as a farmer. A Buffalo Seminole Negro Indian Scout and hero died on March 28, 1928. Pompry Factor was buried in the Seminole Negro Indian scouts Cemetery in Texas.

Lee Fitz
Private, Troop M, Tenth U.S. Cavalry Regiment

Lee Fitz was born in Dinwiddie County, Virginia. He was awarded a Medal of Honor for bravery in action at Tayobaco, Cuba on June 30, 1898. He volunteered to go a shore in the face of the enemy and aided in the rescue of his wounded comrades after several previous attempts had been frustrated. Fitz displayed his unusual feat of bravery during the Spanish American War, 1898.

Henry Ossian Flipper
Second Lieutenant, Tenth U.S. Cavalry Regiment

Henry Ossian Flipper was born in slavery on March 31, 1856, in Thomasville, Georgia. His mother was Isabelle Buckhalter Flipper, (a mulatto) the slave property of the Reverend Reuben Lucky, a Methodist minister. His father, Festus Flipper a skilled bootmaker. He was the slave property of Ephraim G. Ponder, a slave dealer. Henry O. Flipper had several brothers: Joseph, a bishop in the African Methodist Episcopal (AME) Church, Festus, a wealthy farmer, and a brother who was a physician in Jacksonville, Florida. After the Civil War Flipper was appointed to the U.S. Military Academy by James Crawford Freeman, a congressman from Griffin, Georgia (republican representative of the Sixth Congressional District of Georgia). At the academy, Flipper experienced his share of insults and hostility but he faced the loneliness and intimidation with courage and graduated fifteenth in a class of seventy-six on June 14, 1877. Flipper and one of his brothers were taught to read by a slave mechanic who had received permission from his mistress to teach children of her servants. The Flipper brothers were also tutored by an ex-confederate officer's wife for a fee. Henry O. Flipper attended schools operated by the American Missionary Association to include Atlanta University, where he was a member of the first college preparatory class.

Flipper's first military assignment was Fort Sill (Indian Territory) and he was later assigned to Forts Elliot, Texas; Concho, Texas; Quitman, Texas and Davis, Texas. While serving in Indian territory, Flipper was given a message in 1878 to ride his horse for 2 days through hostile Indian Territory to deliver a message concerning the location of Chief Victoria and his band of Indian warriors.

On August 13, 1881, a letter was sent by the commanding officer of Fort Davis, Colonel William Shafter, to the Adjutant General, Department of Texas, accusing Lieutenant Flipper of failing to mail $3,791.77 to the chief commissioner.

The Colonel said that he had seen the lieutenant in town on horseback with saddlebags and he believed that Flipper was leaving the area. Lieutenant Flipper was arrested and even though he had made good all the money for which he was responsible, he was tried by a general court martial at Fort Davis on November 4, 1881. Further investigation revealed that Flipper apparently had not sent the reports and checks to the chief commissary. A number of checks were found. The verdict was not guilty on the charge of embezzlement, but he was found guilty on the charge of conduct unbecoming of an officer and gentleman. His sentence of dismissal from the U.S. Army was confirmed by President Arthur and carried out on June 30, 1882. Efforts were made by several congressional leaders to have his sentence remitted, but unsuccessful, and Flipper resumed civilian life. It has been said that Flipper's enjoyment of horseback riding in the company of a white officer's wife at Fort Concho could have had some effect on the decision of the court martial.

After leaving the Army Henry Flipper worked as a public surveyor and engineer for various mining companies. He was employed by government as an agent to take testimony for land grant cases, and as an engineer for the Sonora Co. Flipper was appointed Deputy U.S. Mineral Surveyor in Nogales, Arizona and worked for Enterprises in Mexico. Henry Flipper had an engineering office in Nogales and at one time was the editor of an all white newspaper, *Nogales Sunday Herald* (1895). His literary accomplishments were: Edited a book on *Mining Laws of Mexico, Negro Frontiersman, The Western Memoirs of Henry O. Flipper, and a translation of Venezuelan Laws, Statutes, and the Colored Cadet at West Point Autobiography of Lieutenant Henry Ossian Flipper.*

Flippers scholarly expertise in the Spanish language was an asset in his professional accomplishments. He served as a chief engineer, and legal advisor to American mining companies in Mexico, Caribbean and South America.

Henry Flipper worked for the Balvanera Mining Co. and Pantepec Oil Co. of New York. He traveled to Spain for the C.B. Ruggles Co. and Texas folklorist J. Frank Dobie to "search for information about a legendary lost Tayopa Mine of Mexico." Flipper was commissioned by the President of Mexico, Porfirio Diaz, to write a three volume history of mining in Mexico.

Henry Flipper possessed superb divergent talents. As surveyor, he surveyed the town of Nogales, Arizona. He was a member of the National Geographical Society, the South West Society, and Archeological Institute of America. He taught physics at Morris Brown College Atlanta, Georgia. Flipper was also an inventor. He was issued a U.S. patient while living in Arizona for inventing a new and useful improvement in tents, a simple light tent that was easy to construct.

Henry Flipper has been recognized through the years for his achievements. The West Point Military Academy unveiled a bronze bust of Flipper. On October 27, 1977 there was an unveiling of a bronze marker at the registered national historic landmark structure at Fort Sill, Oklahoma known as *Flipper's Ditch*: The drainage ditch was an engineering project by Flipper in 1878. The ditch was constructed under the supervision of Flipper with the assistance of soldiers of the Tenth U.S. Cavalry Regiment. The purpose of the ditch was "to eliminate stagnant material ponds and swamps created during the rainy season and to improve the health of the past". The ditch chained north into Medicine Bluff Creek. It is believed that this ditch continues to control flood water and erosion in the specific area at Fort Sill, Oklahoma today.

After completing almost forty years as an outstanding, and competent civil and mining engineer, surveyor, cartographer, inventor, editor, author and a special assistant to the Secretary of Interior, Albert B. Fall, Henry O. Flipper returned to Atlanta, Georgia in 1930 and stayed with his brother Bishop Joseph Flipper of the African Methodist Episcopal Church (AME). In 1922, the Secretary of the Interior, Albert Fall, wrote a letter to the Chairman, Senate Committee on Military Affairs, United States, Senate. Secretary Fall wanted the Senator and his colleagues to consider a bill pending before the committee for the restoration and retirement of Henry Ossian Flipper. *(I have included a copy of the Secretary's letter in Appendix III.)*

The first African American to successfully complete the requirements of the U.S. Military Academy died of a heart attack in Atlanta, Georgia on June 25, 1940 at the age of 84 years.

I personally commend and salute the efforts of a concerned citizen named Ray MacColl in initiating an investigation of the court martial charges with the intentions to clear Flipper of all charges. While he was studying in graduate school at Valdosta State College, Georgia, MacColl, with the assistance of Irsle Fiking and Festus Flipper, filed a formal application with the army board for the correction of military records of Flipper. After 84 years justice was completed for Flipper. The Department of the Army was ordered by the board to correct Flipper's records to show that he was separated from the army of the United States on a certified and honorable discharge on June 30, 1882. The board had found Flipper innocent of the charges of conduct unbecoming of an officer and a gentleman. In December 1976, an army review board awarded him an honorable discharge posthumously. Two years later in 1978, Henry Flipper's body was exhumed from an unmarked grave in Atlanta's Southview Cemetery where he was buried in 1940. After 38 years his body was taken 240 miles from Atlanta to his birth place Thomasville, Georgia.

A ceremony including civilian and military dignitaries and descendants of Henry Flipper were present at his reburial in Thomasville's Magnolia Cemetery. He was buried next to his parents. It was ironic that when they were placing a historic marker at Flipper's gravesite in Georgia, the gravesite of the first black to graduate from West Point, present at the site that day to offer remarks was the first black graduate of the United States Naval Academy in 1949, Commander Wesley Brown. Both men have paved the long broaden road of trials and toil for the eventual triumph and accomplishments of black officers in the military today who attain the coveted ranks of general and admirals. Several years ago I met a white lady from Thomasville, Georgia who said, she and other citizens of Thomasville Georgia are proud to have a famous native come home for peaceful rest.

The memory of Henry Ossin Flipper was revived in 1978 and I believed again in 1994 and I trust for years to come, because he was a true Buffalo Soldier, a veteran of the Indian Wars and above all the Cavalry's first black officer of color. I am in total agreement with the words of the late military historian and archivist, Sara Jackson who wrote:

"Had Henry Ossian Flipper remained in the Army, he would probably have achieved far less. A man of vast strength and strong character, he confronted both prejudices and tradition, and carved out a constructive and productive life for himself."

It is gratifying that today a man of color can look forward to achieving the kind of military career that Lieutenant Flipper might have expected, without experiencing the same difficulties.

George W. Ford
Regimental Quartermaster Sergeant, Tenth U.S. Cavalry Regimental

George Ford was born in 1847 in the state of Virginia. He was the son of Henrietta Bruce and William Ford. His mother was the daughter of Daniel and Hannah Bruce born in 1779 and 1783 in Dumfries, Prince William County, Virginia. They were free people of color. Ford's father was a gardener at Mount Vernon, Virginia. George Ford served ten years in the Tenth U.S. Cavalry Regiment. He enlisted in the Tenth U.S. Cavalry in 1867 and served as sergeant and regimental quartermaster sergeant. Ford was commended in general orders for act of good judgement and gallantry in the Indian wars, and was honorably discharged in 1877. During the Spanish American War, in 1898. Ford enlisted in the Twenty-Third Kansas Infantry Regiment and was commissioned as a Major in the National Guard. After the War, Ford worked as the superintendent of various government cemeteries, including the military cemeteries in Beaufort, South Carolina, Fort Scott, Kansas, Port Hudson, Louisiana, and Springfield, Illinois. He was also appointed treasurer of the Lincoln Exposition in Chicago, Illinois. Ford was active in the Niagara Movement and the National Association For The Advancement of Colored People (NAACP). George Ford was married to Hattie Blythwood. They were the parents of eight children, George William, Noel Bertram, Harriet, James Irvine, Donald, Cecil Bruce, Elsie Gord, and Vera Ford. Their oldest son George W. Ford Jr. had three children, Lena, Hallie and George. Noel Bertram had one daughter, Vera Ford Curtis of Detroit, Michigan. Cecil Bruce Ford, the youngest son had four children. Vera Ford Powell who lived in Germantown, Pennsylvania had two daughters, Ruth Elsie Powell Gaither and Vera L. Powell. Elsie Ford Jenkins who lived in Columbia, South Carolina had one daughter, Carole Douglas Jenkins. George W. Ford Sr. successful military and civilian career and extended family is indicative of the upward mobility trends that were achieved by many "Real Buffalo Soldiers".

John R. Fuller
Private, Ninth U.S. Cavalry Regiment

James R. Fuller enlisted in the Ninth U.S. Cavalry Regiment. He was from Massachusetts, Fuller died on February 12, 1957 and is buried in Arlington National Cemetery.

Nathan Graves
Wagoner, Twenty-fifth Infantry Regiment

Nathan Graves enlisted in the Twenty-fifth Infantry Regiment. He was from the state of Oregon. Graves died on December 18, 1940.

Clinton Greaves
Corporal, Company C, Ninth U.S. Cavalry Regiment

Clinton Greaves was born in 1850 in Madison County, Virginia. He enlisted in the Ninth Cavalry Regiment in 1872 at the age of 22 years. His civilian occupation was a laborer. He was working in Prince Georges County, Maryland, when he entered the army. Greaves was awarded the Medal of Honor on June 26, 1879 for performing gallantry in a hand to hand fight with a band of Chiricahua Apache Indians in the Florida Mountains, New Mexico on June 24, 1877. Greaves retired from military service after 20 years of faithful and courageous service. He retired in Columbus, Ohio. Clinton Greaves a "Real Buffalo Soldier" of the Indian Campaigns died at the age of 56 years, in 1906 at Columbus, Ohio.

Philip Gross
Private, Troop I, Tenth U.S. Cavalry Regiment

Philip Gross enlisted in the Tenth U.S. Cavalry. Gross was a veteran of the Spanish American War. He is buried in Arlington National Cemetery.

Robert G. Gross
Private, Ninth U.S. Cavalry Regiment

Robert G. Gross enlisted in the Ninth U.S. Cavalry. He was from the state of Maryland. Gross died October 18, 1953. He is buried in the Soldiers' Home National Cemetery, Washington, D.C.

Wade H. Hammond
Band Master, Ninth U.S. Cavalry Regiment

Wade Hammond was born in the state of Alabama. He graduated from Alabama Agricultural and Mechanical College in 1895. When he joined the Ninth Cavalry he was assigned as band master. During his military career, he attended the Royal Military School of Music in London. Hammond was presented with a medal by the mayor of Douglas, Arizona. The medal's citation read:

> "In the name of the citizens", "Presented to the Chief Musician, Ninth Cavalry Band, by the citizens of Douglas, Arizona, September 14, 1914. Keep Step To The Music of The Union".

Richard E. Harris
Private, Ninth U.S. Cavalry Regiment

Richard E. Harris enlisted in the Ninth U.S. Cavalry Regiment. He was a veteran of the Spanish American War. Harris was born on July 18, 1874. He died at the age of 81 years on October 11, 1965 and is buried in Arlington National Cemetery.

Willie E. Harris
Private, Troop I, Ninth U.S. Cavalry Regiment

Willie E. Harris was born on May 21, 1878. He enlisted in the Ninth U.S. Cavalry Regiment. Harris served in Cuba during the Spanish American War. Harris died at the age of 63. He is buried in Arlington National Cemetery and "Rest Among The Known".

Arthur Phillip Hayes
Sergeant, Ninth U.S. Cavalry Regiment

Arthur Phillip Hayes was born in Washington, D.C. on February 17, 1896. He was the son of Jackson G. and Maria Tackett Hayes. The successful military and professional career of Hayes demonstrated his desire for upward mobility, and his brief service in the Ninth U.S. Cavalry which also played an integral part in his pursuit of excellence in his endeavors. While pursuing his goals, he enlisted in the Ninth U.S. Cavalry regiment in 1919 and was promoted to the rank of Sergeant. During World War I, he studied at Howard University, Washington, D.C. and then entered the military service as a private and served briefly on active duty as a second lieutenant. Hayes received a Bachelor of Science degree in 1927 from Lincoln University of Missouri, and a Master degree from Columbia University, New York in 1932.

In 1922, Arthur P. Hayes received a commission as a lieutenant in the Infantry Reserve. He served as an assistant professor of Military Science and Tactics at Lincoln University of Missouri, 1924-1927 and Prairie View Agricultural and Mechanical College, Texas, 1930-1931. During World War II, Hayes served in the rank of Major in the Army Air Corps at Tuskegee Alabama, Selfridge Field, Michigan, Paterson Field, Indiana, Godman Field, Kentucky and Freedman Field, Indiana. In 1949, Arthur P. Hayes was a Lieutenant Colonel, New York National Guard, 176th Military Police. The military career of Arthur P. Hayes was commendable and it is significant that he served a few years during his upward mobility strides in the Ninth U.S. Cavalry Regiment.

Samuel T. Hodge
Private, Ninth U.S. Cavalry Regiment

Samuel T. Hodge enlisted in the Ninth Cavalry Regiment. He was from the state of Maryland. Hodge died on February 4, 1950 and "Rest Among The Known" in Arlington National Cemetery.

Emmett G. Jackson
First Sergeant, Tenth U.S. Cavalry Regiment

Emmett Jackson was a member of the Tenth U.S. Cavalry. He was married to Beatrice Ware Jackson who was born on November 9, 1894. Emmett Jackson died on August 29, 1944. His wife Beatrice died at the age of 86 years on February 25, 1980. Emmett and Beatrice Jackson are buried in Arlington National Cemetery.

John Jackson
Private, Tenth U.S. Cavalry Regiment

John Jackson enlisted in the Tenth U.S. Cavalry. He was from the District of Columbia. Jackson died September 26, 1929. He is buried in the Soldiers' Home National Cemetery, Washington, D.C.

Lewis A. Jackson
Private, Troop D, Tenth U.S. Cavalry Regiment

Lewis A. Jackson enlisted in the Tenth U.S. Cavalry Regiment. He was born on July 6, 1880. Jackson was married to Landonia C. Jackson who was born on October 4, 1879. Lewis Jackson died at the age of 81 years on May 23, 1961. His wife Landonia died on February 18, 1946. Lewis and Landonia Jackson are buried in Arlington National Cemetery.

Lon Jackson
Wagoner, Tenth U.S. Cavalry Regiment

Lon Jackson enlisted in the Tenth U.S. Cavalry Regiment. He was born on February 16, 1876. Jackson died at the age of 74 years on May 5, 1950. He is buried in Arlington National Cemetery.

Andrew W. Johnson
Private, Troop K, Ninth U.S. Cavalry Regiment

Andrew W. Johnson was born in October 1895. He enlisted in the Ninth U.S. Cavalry Regiment. He was married to Hannah R. Johnson who was born on March 21, 1881 and died June 29, 1964. Andrew Johnson died on June 10, 1938. Private Andrew Johnson and his wife Hannah are buried in Arlington National Cemetery.

Elias Johnson
Musician, Company F, Twenty-fifth Infantry Regiment

Elias Johnson was born August 16, 1870. He enlisted in the Twenty-fifth Infantry Regiment. Johnson served during the Spanish American War. He was married to Lillian Johnson who was born on February 12, 1878. Elias and Lillian Johnson are buried in Arlington National Cemetery. Elias died August 8, 1938.

Henry Johnson
Sergeant, Troop D, Ninth U.S. Cavalry Regiment

Henry Johnson was a native of Boynton, Virginia. He enlisted in the Ninth Cavalry at the age of 21 years in 1867. His civilian occupation was a laborer. During the period, October 2-5, 1879, Johnson displayed unusual valor for which he was awarded the Medal of Honor on September 22, 1890. Sergeant Johnson was on a rescue mission with his Troop or company when they were sent to Milk River, Colorado to provide assistance and relief to

200 white soldiers who were under attack by the Ute Indians from the White River Agency. Johnson demonstrated his unusual valor when he voluntarily left the fortified shelter and under heavy fire at close range made the rounds of the pits to give instructions to the guards. He fought his way to a nearby creek and obtained water for the wounded soldiers. Henry Johnson retired around 1900 and lived in Washington, D.C. until his death in 1904. He is buried in Arlington National Cemetery.

Joseph H. Johnson
Trooper, Ninth U.S. Cavalry

Joseph Johnson was born in Washington, D.C. Before joining the army, he worked on the railroad, but in 1880 he enlisted in the Ninth U.S. Cavalry. His performance was outstanding. He won several sharp shooter medals. In 1891, Johnson joined the National Guard as a private and was promoted to first sergeant and later, in 1892, adjutant. When the Eight Illinois Regiment was mustered into service in the Spanish American War, a former Buffalo Soldier was commissioned lieutenant colonel. He served with the regiment in Cuba where he held the distinctive position of senior field grade field.

Mose Johnston
Sergeant, Machine Gun Troop, Tenth U.S. Cavalry Regiment

Mose Johnston was born on February 8, 1885 and was from the state of New York. He served in World War I. Johnston died at the age of 63 years and is buried in Arlington National Cemetery where he "Rest Among The Known"

Walter C. Jones
Private, Ninth U.S. Cavalry Regiment

William C. Jones enlisted in the Ninth U.S. Cavalry Regiment. He was from Washington, D.C. Jones was married to Emily Jones who was born on January 1, 1877. William Jones died November 8, 1938. His wife Emily died at the age of 75 years on September 13, 1952. Walter and Emily Jones are buried in Arlington National Cemetery.

George Jordan
Sergeant, Troop K, Ninth U.S. Cavalry Regiment

George Jordan was born in Williamson County, Tennessee. He enlisted in the army at the age of 19 years in 1866 in Nashville, Tennessee. He was assigned to Troop K, Ninth U.S. Cavalry Regiment. He twice exhibited unusual heroism. On May 14, 1880, while commanding a detachment of twenty-five men at Tularosa, New Mexico, he repulsed a force of more than one hundred Indians at Carrizo Canyon. In the following year on August 12, 1881, while commanding a detachment of 19 men, he stubbornly held his ground in an extremely exposed position and gallantly forced back a much superior number of Indians, preventing them from surrounding the command. Jordan and his men were responsible for protecting the white citizens of Tularosa from a fierce Indian attack. Sergeant George Jordan died in 1904 in Crawford, Nebraska and is buried in the post cemetery at Fort Robinson.

Daniel Kelly
Private, Tenth U.S. Cavalry Regiment

Daniel Kelly was a member of the Tenth U.S. Cavalry. He was from the State of Pennsylvania. Kelley died September 19, 1923. He is buried in the Soldiers' Home National Cemetery, Washington, D.C.

Isaac Landon
Private, Ninth U.S. Cavalry Regiment

Isaac Landon was a member of the Ninth U.S. Cavalry. He was from the state of Virginia. Landon died April 8, 1925. He responded to the "call of taps" and is buried in the Soldiers' Home National Cemetery, Washington, D.C.

Frank Lewis
Private, Ninth and Tenth U.S. Cavalry Regiments

Frank Lewis was born in 1849 and in 1907 was living in San Antonio, Texas. Lewis served during the Civil War in the United States Colored Troops (USCT). After the Civil War he reenlisted on September 22, 1866 in Company B, Ninth U.S. Cavalry and served for seven years. In 1873, Burton enlisted in Company L, Tenth U.S. Cavalry and served for ten years and was discharged on October 8, 1883. He was receiving a pension of eight dollars a month in 1907. Burton was stationed at Fort Stockton during some of his military career. He worked as a custodian at the San Antonio Texas City Hall in 1907. Frank Lewis served during the Indian Campaign.

Charles Logan
Sergeant, Company K, Twenty-fifth Infantry Regiment

Charles Logan was a member of the Twenty-fifth Infantry. Logan served in World War I. He died January 20, 1918. Logan is buried in the Soldiers' Home National Cemetery, Washington, D.C.

John A. Logan
Sergeant Major, Ninth U.S. Cavalry Regiment

John Logan enlisted in the army on August 21, 1892 at Chattanooga, Tennessee, and was assigned to Troop C, Ninth U.S. Cavalry. He served for five years with Troop C and six with Troop L of the same regiment, and was promoted to Sergeant Major on August 29, 1904. In 1907, he was awarded the title of Marksman for his prowess with weapons. Logan served for some time in the Philippines and fought at the battle of La Quasima, Cuba, July 1 to 11, 1898.

Frank W. Love
Sergeant Major, Twenty-fifth Infantry Regiment and Ninth U.S. Cavalry

Frank Love was born in Kansas City, Missouri on November 20, 1877. He enlisted in the army on August 18, 1900, and was assigned to Company D, Twenty-fifth Infantry Regiment. Love served in the Philippine Islands from October 31, 1900 to July 8, 1902 and from May 31, 1907 to May 15, 1909. Frank Love was assigned to the Ninth U.S. Cavalry Regiment in August 1906. He was appointed Squadron Sergeant Major on January 16, 1909.

Walter H. Loving
Captain, Bandmaster, Twenty-fourth Infantry Regiment

Walter Loving was born in Lovington, Nelson County, Virginia on December 17, 1872, and attended Dunbar High School (the old M Street school). He enlisted in the Twenty-fourth U.S. Infantry Regiment, and in 1899 served as bandmaster of the Eighth Illinois Volunteers. (He completed a special course in band conducting at the New England Conservatory.) Loving was commissioned second lieutenant and also served with the Forty-Eighth U.S. Volunteers in the Philippines, where on February 13, 1902, he was assigned the task of organizing a band for the Philippine Constabulary Service. Captain Loving participated in the St. Louis and Seattle expositions in 1906. His performance as a musician was commendable, a credit to the military and a testament to the Negro's progress during that period.

Gurnzee M. Lucas
Corporal, Ninth U.S. Cavalry Regiment

Gurnzee M. Lucas enlisted in the Ninth Cavalry Regiment. He was a member of the Regimental Band. Lucas was married to Maude Lucas who was born on December 15, 1895. Corporal Lucas died on March 14, 1932. His wife Maude died on January 17, 1990. Gurnzee and Maude Lucas are buried in Arlington National Cemetery.

Manuel Mack
Private, Ninth U.S. Cavalry Regiment

Manuel Mack enlisted in the Tenth U.S. Cavalry Regiment. He was born on January 19, 1886. Mack served in Cuba during the Spanish American War. He was married to Grace B. Mack who was born on October 27, 1881. Manuel Mack died on April 18, 1961. Manuel and Grace B. Mack are buried in Arlington National Cemetery.

Patrick Mason Jr. Sergeant,
Twenty-fourth and Twenty-fifth United States Infantry Regiments

Patrick Mason Jr. was born in 1844 in Virginia near Halifax. His parents were Patrick Sr. and Catherine Delany Mason. His family moved to Oberlin, Ohio in the 1850's. Patrick's father was a wagon maker and his mother worked as a domestic. His family consisted of eleven children. Some of Patrick's brothers and sisters graduated from Oberlin College Ohio.

Patrick enlisted at the age of 26 in 1870 and served for five years in the Twenty-fifth United States Infantry and his remaining career was served in the Twenty-fourth Infantry Regiments.

Patrick Mason Jr. was stationed at Camp Supply in Indian Territory, (Oklahoma) Fort Bayard, New Mexico, Fort Douglas, Utah. Patrick married Victoria Rivers, daughter of Augustus and Mary Rivers in 1886. Their only child was born in June 1890 at Fort Bayard.

Victoria Mason died shortly after the birth of her daughter and Patrick had Mary and Augustus Rivers to raise his child.

Sergeant Patrick Mason Jr. served in Cuba during the Spanish American War and later was stationed at Fort D.A. Russell, Wyoming. Mason died of typhoid fever at Cabanatuan, Island of Luzon, Philippines.

George Marshall
Cook, Twenty-fifth Infantry Regiment

George Marshall enlisted in the Twenty-fifth Infantry Regiment. He was from the state of Missouri. Marshall died on January 11, 1929. He is buried in Arlington National Cemetery and "Rest Among The Known".

Hezekiah Matthews
Private, Tenth U.S. Cavalry Regiment

Hezekiah Matthews was a native of Washington, D.C. He was a member of the Tenth U.S. Cavalry Regiment. Matthews died on March 11, 1940.

Isaiah Mays
Corporal, Company B, Twenty-fourth, U.S. Infantry Regiment

Isaiah Mays was born in 1858 in Carters Bridge, Virginia. At the age of 23, he enlisted in the Twenty-fourth Infantry Regiment. His civilian occupation was a laborer. Mays was awarded the Medal of Honor for gallantry in a fight between Paymaster Whams' escorts of a $28,345.10 payroll on May 11, 1889 and the civilian outlaws. It has been stated that Corporal Mays "walked and crawled two miles to Cottonwood Ranch to get relief". Mays was discharged from the army in 1893 and resided in Arizona and New Mexico. Isaiah Mays was a Buffalo Soldier of the Twenty-fourth Infantry Regiment who demonstrated courage and initiative while defending Major Joseph W. Wham's payroll stage coach.

Thomas McBride
Private, Ninth U.S. Cavalry Regiment

Thomas McBride enlisted in the Ninth U.S. Cavalry Regiment. He was married to Mary McBride who was born on October 28, 1894 and died at the age of 66 years on November 25, 1960. Thomas and Mary McBride are buried in Arlington National Cemetery.

Tom McBride
Private, Ninth U.S. Cavalry Regiment

Tom McBride enlisted in the Ninth U.S. Cavalry. He was a native of Washington, D.C. McBride died on April 2, 1942. Private McBride "Rests Among The Known" at Arlington National Cemetery.

William McBryar
Sergeant, Company K, Tenth U.S. Cavalry Regiment

Sergeant William McBryar was born in 1861 in Elizabeth, North Carolina. He enlisted in the Tenth Cavalry Regiment in 1887 in New York City at the age of 25 years. His civilian occupation was a laborer. His Medal of Honor citation reads *"Sergeant McBryar showed outstanding bravery in battle with Apache Indians"*. He was awarded the Medal of Honor on May 15, 1890. McBryar was with a detachment of Buffalo Soldiers under the command of Lieutenants Powhattan H. Clark and James Watson. The detachment along with other units had been alerted to pursue a band of Apache Indians who had killed a freight wagon driver near Fort Thomas. McBryar's commanding officer, Lieutenant Watson stated that William McBryar *"distinguished himself for coolness, bravery and good marksmanship under circumstances very different from those on the target range."*

During the Spanish American War - 1898, McBryar received a commission as a first lieutenant in the Eighth United States Volunteer Infantry. Sergeant William McBryar was discharged around 1901. He was interested in pursuing excellence in education and in his seventies he attended the Tennessee State (University today) Agricultural and Industrial College in Nashville, Tennessee. McBryar worked in various job positions during his civilian career. He worked as "a watchman at Arlington National Cemetery, truck farmer, guard at the McNeil Island, Washington Federal Penitentiary, and a private and public school teacher". William McBryar died in Philadelphia in 1941. He was eighty years of age. The distinguished Medal of Honor recipient and Buffalo Soldier who attained numerous successful goals in civilian and military life is buried in Arlington National Cemetery.

Martin Mc Dowell
Staff Sergeant, Ninth and Tenth Cavalry Regiments

Martin McDowell was born on March 3, 1879 in Louisville, Kentucky. His parents were Henry McDowell, a circuit minister and Annie McDowell. McDowell enlisted in the Ninth Cavalry in 1907 at Fort Leavenworth, Kansas. He served in the Philippine Islands and Fort D.A. Russell, Wyoming. McDowell was discharged in 1911. Martin McDowell decided to reenlist. On January 3, 1911, he reenlisted in the Army Service School at Fort Leavenworth, Kansas where he was assigned to the Tenth U.S. Cavalry. McDowell demonstrated a commendable proficiency in riding techniques and the knowledge of the training. McDowell married Lillian Banks of Laton, Oklahoma and they were the parents of a son. Staff Sergeant McDowell had some young admirers while serving at Fort Leavenworth, Kansas These admirers were young girls, children of the officers assigned at Fort Leavenworth and from families outside of the Fort.

The girls had formed a Fort Leavenworth Riding club and participated in the Annual American Royal Exhibition Riding Stock Shows. Their mentor and instructor was Sergeant McDowell. The young girls had riding class twice weekly. McDowell coached the young ladies on riding drills and techniques. They learned how to ride correctly, jump and perform general horsemanship.

Staff Sergeant Martin McDowell was regarded very highly by the young ladies and their parents. McDowell served over 24 years in the U.S. Army.

Hugh Mc Elroy
Private, Tenth U.S. Cavalry Regiment

Hugh McElroy served in the Spanish American War and was with the Tenth U.S. Cavalry Regiment in the Punitive Expedition to Mexico in 1916. During World War I, McElroy served with the three hundred and seventeenth engineers, but was assigned temporarily to the French Seventh Army. While serving with the French, he was awarded the Croix de Guerre for gallantry in action.

William McLain
Private, Company H, Twenty-fourth Infantry Regiment

William McLain was born on January 1, 1894. He enlisted in the Twenty-fourth Infantry Regiment. McLain served during World War I. He died on May 5, 1943 at the age of 49 years. William McLain is buried in Arlington National Cemetery where he "Rest Among The Known."

Jerry Milliman
Trooper, Tenth U.S. Cavalry Regiment

Jerry Milliman, alias Jeremiah, was born in Glen Falls, New York in 1840. He enlisted on April 9, 1863 for three years at Readville, Massachusetts by R.O. Hallowell. The official records listed his description as 23 years, five feet, five inches, dark complexion, black eyes and black hair. His civilian occupation was a boatman and farmer. Milliman was present for duty at the skirmish on James Island, South Carolina, July 16, 1863; the Battle of Honey Hill, South Carolina, November 30, 1864 and Fort Wagner, Morris Island, South Carolina July 18, 1863. He was promoted to Corporal on June 22, 1863. Milliman was discharged on August 20, 1865.

After the Civil War, Milliman decided to enlist in one of the regular United States Army units that were activated. He enlisted on January 4, 1869 and was assigned to Troop L, United States Army Tenth Cavalry. Milliman served as a Farrier and was discharged from the regular U.S. army on December 2, 1878. He had served at the following posts; Fort Sill, Oklahoma; Fort Concho near Angelo, Texas; and Fort Stanton.

Jerry Milliman was married twice. His first wife was Rosa Ann Jackson. She died in March, 1878. Milliman's second wife was Elizabeth Van. They were married on November 30, 1879 in San Antonio, Texas by the Reverend G.W. Townsend. The witnesses present were T.J. and A.LA. McClery.

Milliman died on December 27, 1896. His widow, Elizabeth applied for a widow's pension and it was necessary for her to provide detailed information concerning any previous marriages. The following information depicts some interesting experiences that former slaves confronted during the institution of slavery. Elizabeth Van Milliman made the following statements in a second affidavit in support of her pension claim: *my former master, Seth Randall brought me to Texas from Talladega, Alabama. I was called Rachel, however my full name is Rachel Elizabeth. After Emancipation, I used the name Bettie Van because Van was my father's name. During slavery, I had one husband. His name was Henry Randall. We belonged to Seth Randall. A white preacher performed the marriage ceremony.*

Henry Milner
Private, Company K, Twenty-fourth Infantry Regiment

Henry Milner enlisted in the Twenty-fourth Infantry Regiment. He served in Cuba during the Spanish American War. Milner died on July 26, 1933. He is buried in Arlington National Cemetery and answered to the "final call of taps".

Miles Moore
Musician, Twenty-fifth Infantry Regiment

Miles Moore was named David Miles Moore. However, he used the name Miles Moore. He was born on April 8, 1848 in Ithaca, New York. He enlisted on April 29, 1863 at Readville, Massachusetts. He was described as five feet, ten inches, black hair mixed with gray, brown eyes and mulatto complexion. He was discharged on August 30, 1865. Moore's parents were David and Elizabeth Moore. His father was born in Elmira, New York.

Moore decided to continue his military service after the Civil War. He served in the following units after 1865. Enlisted in Company E, Thirty-ninth United States Colored Infantry (USCI) on May 13, 1868. He was at Fort Columbus, New York where he was a musician in the band. On April 30, 1870, Miles was transferred to Company F, Twenty-fifth U.S. Infantry and was a musician in the band. (The Thirty-ninth and Fortieth USCT Regiments were consolidated into the Twenty-fifth Regiment USCT). Miles was discharged on August 30, 1870 at Fort Clark, Texas.

Miles married Ardelle Rosemary on December 16, 1875. They were married by Rev. Stephen Priestly at New Orleans, Louisiana. They were the parents of Richard, born April 6, 1874; Augustus, born October 22, 1877; Arthur, born January 12, 1883; Elizabeth, born October 12, 1888 and George, born March 9, 1890.

David Miles Moore died on May 30, 1904 at the hospital in Saratoga Springs, New York. He lived at 63 Walwork Street, Saratoga Springs, New York.

Charles S. Murray
Private, Troop I, Ninth U.S. Cavalry Regiment

Charles S. Murray was a member of the Ninth U.S. Cavalry. Murray was from the state of Virginia. He was born July 4, 1884 and died February 20, 1960. Murray is buried in the Soldiers' Home National Cemetery, Washington, D.C.

Adam Paine
Private, Detachment of Seminole Scouts U.S. Army

Adam Paine was born in Florida. He enlisted in the army on November 12, 1873 at Fort Duncan, Texas. Paine was awarded the Medal of Honor on October 13, 1875 for gallantry in action on September 20, 1874 at Staked Plains, Texas. Adam Paine displayed his gallant performances during the Red River War (1874-1875) involving an engagement at Palo Duro Canyon, with Comanche Indian Warriors. Paine was discharged from the army on February 19, 1875. Later, he moved to Brownsville, Texas and Brackettville, Texas areas. He was shot by a sheriff on January 1, 1877 because he was considered a fugitive from justice.

Alfred J. Patton
Private, Company C Twenty-fourth Infantry Regiment

Alfred J. Patton enlisted in the Twenty-fourth Infantry Regiment. Patton was born on October 26, 1894. He died in 1946 and is buried in Arlington National Cemetery.

Isaac Payne
Private (Trumpeter) Detachment of Seminole Negro Indian Scouts, U.S. Army

Isaac Payne was born in Mexico. He enlisted in the Army at the age of seventeen on October 7, 1871 at Fort Duncan, Texas and served until his discharge on January 21, 1901 at Fort Ringold, Texas. He was discharged but then reenlisted continuing his period of military service. He received the Medal of Honor for helping Lieutenant Bullis to escape an approaching party of Indians at Pecos River, Texas on April 25, 1875. He was recommended for the Medal of Honor along with two other Seminole Negro Indian scouts, Private Pompey Factor and Sergeant John Ward. Isaac Payne lived at Seminole Negro Indian Camp when he retired. Payne died on January 12, 1904 at Bacunuebtim - Nacimiento - Coahuila, Mexico.

William H. Penn
Sergeant, Third Squadron, Ninth U.S. Cavalry Regiment

William Penn was born in 1863 at Baltimore, Maryland. He enlisted in the Army before his seventeenth birthday and was assigned to the Ninth Cavalry. He served in the Indian Wars and in Cuba, the Philippines and Samoa Islands. It is believed that Penn's father and two Uncles were killed in the Civil War. Penn retired from military service on February 14, 1908.

Beverly Perea
First Lieutenant, Twenty-fourth Infantry Regiment

Beverly Perea was born in Mecklenburg, Virginia. He and his wife, Missouri, had one daughter. Perea enlisted in the army on July 25, 1871, and was assigned at various times to companies A, I, E, B and M of the Twenty-fourth Infantry Regiment. Perea served continuously for thirty-one years. He was two years, four months and sixteen days on foreign service in the Philippines and for two months and seven days in Cuba. He was cited as a first classman in marksmanship in 1897. Perea was appointed second lieutenant, Seventh U.S. Volunteers on September 16, 1898, first lieutenant on January 7, 1899, and provost marshal as an additional duty in 1901.

According to Special Orders 41, Headquarters, 2nd District, Department of Northern Luzon, Apani, Philippines Island, dated February 19, 1901, "Second Lieutenant Perea, Forty-Ninth Infantry USV, was detailed additional duty as provost marshal at Pamplona, P.I." Lieutenant Perea died on April 3, 1915, and was the first known Negro officer to be buried with honors in the National Military Cemetery at Arlington, which had been his dying wish.

Several newspapers carried the account of his death. A Boston paper published in April, 1915, read as follows:

In response to an appeal of the widow of First Lieutenant Beverly Perea, U.S.A., retired, a colored citizen who died at the Cambridge Hospital Saturday, Mayor Curley has requested Secretary of War Lindley M. Garrison to give permission for the interment of the lieutenant's remains in the Arlington National Cemetery at Washington.

Major Curley's personal appeal to Secretary Garrison was successful and Lieutenant Perea was buried at Arlington Cemetery on April 10, 1915.

John Pierce
Private, Troop M, Tenth U.S. Cavalry Regiment

John Pierce was a member of the Tenth U.S. Cavalry Regiment. Pierce served in the Spanish American War. He is buried in Arlington National Cemetery.

Henry Vinton Plummer
Chaplain, Ninth Cavalry Regiment

Henry Vinton Plummer was born a slave June 30, 1844 in Prince Georges County, Maryland. During the Civil War he enlisted in the Navy for sixteen months, and was discharged in August 1864. After his discharge, Plummer attended Wayland Seminary, Washington, D.C. and graduated in 1879. He served as a baptist missionary in Maryland.

On July 1, 1884, Plummer became the first black man commissioned in the regular Army as a chaplain. His first assignment was Fort Riley, Kansas and later he served at Fort McKinney, Wyoming Territory and Fort Robinson, Nebraska.

Plummer's duties as a chaplain were to conduct Sunday worships, Sunday school and supervise the church choir activities. Plummer supported the Temperance Movement and he gave lectures on temperance and started a "Loyal Temperance Legion" at Fort Robinson, Nebraska. His early performances of duty were regarded by his superior as "good in professional ability, general conduct, discipline and habits." The post commander enlisted men, and some officers were very pleased with Chaplain Plummer's Sunday services and religious programs.

Plummer was a member of the chaplain's movement whose goals were to increase the number of chaplains and to interest the churches and congress to support the recruitment of qualified and dedicated chaplains for the regular U.S. Army. Plummer was against the sale of alcoholic beverages and wine in the Post Canteen. The military authorities were not in complete agreement with Plummer's ideas and plans to prohibit the sale of alcoholic beverages on military posts. Chaplain Plummer wrote a Weekly Post Bulletin about the town of Crawford, Nebraska and citizen race relations with the black soldiers. He also suggested that the black soldiers should boycott some of the saloons and businesses in the town. The Post Commander did not appreciate these actions of Plummer. Plummer also wanted to visit Central Africa with some black troops on a missionary tour. He demonstrated a very sincere and strong Nationalistic belief in assisting the African people.

On April 20, 1884, Plummer requested through military channels that he be permitted to visit Africa with approximately 100 soldiers from the Ninth, Tenth, Twenty-fourth and Twenty-fifth regiments. Plummer wanted to "introduce American civilization and christianity among some of the African tribes." His plans were supported by a former black chaplain who served in the Civil War, Bishop Henry McNeal Turner, African Methodist Episcopal (AME) Church and other black religious leaders.

Unfortunately in June 1884, Chaplain Plummer was experiencing some serious difficulties in his military career when he was accused of some negative moral actions. Some enlisted men had complained that Chaplain Plummer had been engaged in drinking intoxicating liquor with the enlisted men and had given liquor to enlisted men. He was also accused of being intoxicated in the presence of a Sergeant and his wife.

The military authorities did not permit Chaplain Plummer to voluntarily resign because of these allegations against him. A decision was made to court martial Plummer. There were some black leaders who wrote letters in support of Chaplain Plummer. Bishop Henry McNeal Turner wrote President Grover Cleveland on October 18, 1884 and asked for clemency and not dismissal from active military service for Plummer. Congressman John Mercer Langston, from Virginia wrote President Cleveland and enclosed letters, affidavits and testimonies from people stating positive remarks about Plummer's character and sobriety.

President Cleveland decided to approve the final actions of Plummer's court martial on November 2, 1894. On November 10, 1894, the first regular Army black Chaplain was dismissed from active military duty.

Chaplain Henry V. Plummer and his wife and four children moved to Kansas City, Kansas. Plummer served as a pastor in churches in Wichita and Kansas City, Kansas. A former Buffalo Soldier, a concerned chaplain and a man aware of his ancestral roots, Africa died on February 10, 1905 in Kansas City, Kansas.

Peter Pogue
Corporal Twenty-fifth Infantry Regiment

Peter Pogue was born on July 10, 1875. He enlisted in the Twenty-fourth Infantry Regiment and was a member of the regiment's band. Pogue was married to Erma (Einnis) Pogue. She was born on March 2, 1880 and died May 25, 1943. Peter and Erma Pogue are buried in Arlington National Cemetery.

Edward Polk
Master Sergeant, Twenty-fourth Infantry Regiment

Edward Polk was born on November 23, 1869. He enlisted in the Twenty-fourth Infantry Regiment. Polk was a member of the Twenty-fourth Infantry band. He was from the state of Mississippi. Polk who was born on June 30, 1884 died at the age of 80 years on August 18, 1964. Master Sergeant Edward Polk and his wife Nannie are buried in Arlington National Cemetery.

Emment Preston
Private, Tenth U.S. Cavalry Regiment

Emment Preston enlisted in the Tenth U.S. Cavalry Regiment. He was born on July 9, 1873. Preston died at the age of 85 years on January 15, 1959. He is buried in Arlington National Cemetery.

George W. Prioleau
Chaplain, Ninth U.S. Cavalry

George Prioleau was born in Charleston, South Carolina. He attended public schools in Charleston, the Avery Institute and Clafin University. He taught in public schools for a time and then entered the ministry. After graduating from the Theological School of Wilberforce University in 1884, he pastored several churches in Xenia, Ohio. In 1895, he was appointed chaplain of the Ninth U.S. Cavalry Regiment, with the rank of Captain and served until 1915, when he transferred to the Tenth and finally to the Twenty-fifth Infantry Regiment. He served at Fort Missoula, Montana in 1891 and the Philippines. George W. Prioleau was a Methodist minister.

Howard Donovan Queen
Trooper, Tenth U.S. Cavalry Regiment

Howard Queen was born on November 18, 1894, in Tee Bee, Maryland, the son of Richard Thomas and Rebecca Virginia Queen. He received a degree in engineering from Howard University and also attended various military schools before embarking on an outstanding military career.

Queen enlisted in the Washington, D.C., National Guard in 1910, and served in the Tenth Cavalry from 1911 to 1917. In 1913, he was a corporal with his regiment in Winchester, Virginia, and was duly promoted until he reached the rank of colonel. In 1941, he became the commanding officer of the Three Hundred and Sixty-Sixth Infantry Regiment.

Colonel Queens' tactical experience included the battle at El Carrizal, Mexico, on June 21, 1916; the Punitive Expedition to Mexico with the Tenth Cavalry from 1916-1917; and service in France in the Vosges sector with the Three Hundred and Sixty-Eighth Infantry, the Meuse-Argonne offensive and the Metz sector from 1918-1919, and operations with the Three Hundred and Sixty-Sixth Infantry Regiment. His military lineage can be traced back to the American Revolution. His great-grandfather was present at the Battle of Boston Common; his grandfather and three uncles served with the Union troops during the Civil War, and his father served for fifteen years with the Twenty-fourth Infantry Regiment and the Tenth U.S. Cavalry. His cousin lost his life aboard the USS Maine in the Spanish-American War, and a brother served as an officer with the Three Hundred and Sixty-Eighth Infantry in the Argonne offensive.

John T. Ray
Private, Troop G, Tenth U.S. Cavalry Regiment

John T. Ray was born on January 2, 1871. He enlisted in the Tenth Cavalry Regiment. Ray died in July 1954 and is buried in Arlington National Cemetery where he responded to the "call of taps".

Ernest Reed
Private, Ninth U.S. Cavalry Regiment

Ernest Reed enlisted in the Ninth U.S. Cavalry Regiment. He was from Connecticut. Reed died on February 28, 1931 and is buried in Arlington National Cemetery.

George Reid
Private, Company M, Twenty-fourth Infantry Regiment

George Reid enlisted in the Twenty-fourth Infantry Regiment. He served in Cuba during the Spanish American War. Reid was married to Julia Reid who was born on April 10, 1876. Reid died on November 17, 1946. His wife Julia died on September 5, 1956. George Reid and Julia Reid are buried in Arlington National Cemetery.

Robert Reynolds
Sergeant, Troop B, Tenth U.S. Cavalry Regiment

Robert Reynolds was born on March 6, 1874. He was from the state of Maryland. Reynolds was married to Sophia Reynolds who was born on January 20, 1872. Robert Reynolds died on January 19, 1923. Reynolds and his wife Sophia are buried in Arlington National Cemetery.

Robert P. Rhea
Corporal, Troop M, Ninth U.S. Cavalry Regiment

Robert P. Rhea was born on February 17, 1874. He enlisted in the Ninth Cavalry Regiment. Rhea served in Cuba during the Spanish American War. He was married to Annie Rhea who was born on June 10, 1876 and died March 31, 1962. Robert Rhea died June 10, 1938. Corporal Robert P. Rhea and his wife Annie are buried in Arlington National Cemetery.

Marion Colas Rhoten
First Lieutenant, Ninth and Tenth Cavalry Regiments

Marion Colas Rhoten was born in Fayetteville, Tennessee on January 1, 1877. He was the son of Charles and Katy Wagner Rhoten. He married Celia B. Tabor of Texas. They were the parents of six children, Morris L, Wilma L, Eugene, Philip, Clifford and Marion A. Marion C. Rhoten enlisted in the U.S. Army in 1910. He served in the Ninth and Tenth Cavalry regiments. Rhoten was stationed at Mounted Service School Fort Riley, Kansas. During World War I, he attended Fort Des Moines, Iowa and was commissioned a First Lieutenant. Rhoten trained troops for overseas duty at Camp Funston. Later he served overseas with the 92nd Division. After the War, he moved to Albuquerque, New Mexico and operated the M.C. Rhoten Real Estate agency. He was a Grand Master of the Prince Hall Masons. Marion C. Rhoten was a successful business man and his early service as a Buffalo Soldier provided him an opportunity to continue to pursue excellence in his endeavors.

Philip Robb
Private, Tenth U.S. Cavalry

Philip Robb was a member of Tenth Cavalry Regiment. He was born on October 6, 1880. Robb was from Washington, D.C. Philip Robb died at the age of 66. He is buried at Arlington National Cemetery.

Aaron Robinson
Private, Tenth U.S. Cavalry Regiment

Aaron Robinson was a member of the Tenth U.S. Cavalry. Robinson died October 18, 1929. He is buried in the Soldiers' Home National Cemetery, Washington, D.C.

Benjamin D. Robinson
Private, Ninth U.S. Cavalry Regiment

Benjamin D. Robinson enlisted in the Ninth U.S. Cavalry. Robinson died in September, 1950. He is buried in the Soldiers' Home National Cemetery, Washington, D.C.

John C. Sanders
First Sergeant, Tenth U.S. Cavalry Regiment

John Sanders served with the Tenth U.S. Cavalry Regiment for twenty-nine years, seven months and fifteen days. At the time of his retirement on February 3, 1942, Sanders had accumulated the longest service record of any first sergeant in the army, having been promoted to that grade in 1916.

Robert Sayles
Cook, Troop B, Ninth U.S. Cavalry Regiment

Robert Sayles enlisted in the Ninth Cavalry Regiment. He was born on June 25, 1877. Sayles was married to Ada V. Sayles. Robert Sayles died at the age of 70 years on November 28, 1947. His wife was born on June 13, 1884 and died at the age of 74 years on February 1, 1958. Robert and Ada Sayles are buried in Arlington National Cemetery.

Charles Scott
Private, Twenty-fifth Infantry Regiment

Charles Scott enlisted in the Twenty-fifth Infantry Regiment. Scott is buried in Arlington National Cemetery. He died on August 12, 1936.

Oscar J. W. Scott
Chaplain, Twenty-fifth U.S. Infantry Regiment

Oscar J. W. Scott was born in Gallipolis, Ohio, on July 31, 1867. He was educated in the Ohio public schools. Scott attended Ohio's Wesleyan University and Drew Theological Seminary. He was a minister in the African Methodist Episcopal Church and a former pastor of the Metropolitan AME Church, Washington, D.C. Chaplain Scott was appointed Chaplain in the U.S. Army on April 17, 1907. He served with the Twenty-fifth Infantry Regiment in Texas, in the state of Washington and overseas in the Philippines and Hawaii. Scott was married to Nettie Poindexter. He died on March 13, 1928. Chaplain Oscar J. W. Scott is buried in Arlington National Cemetery.

Thomas Shaw
Sergeant Troop B, Ninth U.S. Cavalry Regiment

Thomas Shaw was born a slave in 1846 in Covington, Kentucky. He was the slave property of Mary J. Shaw, a Pike County Missouri citizen. When Thomas Shaw decided to enlist in Company A, 65th Colored Infantry during the Civil War, Mary Pike became concerned about her 18 year old field hand slave. She filed a claim for compensation for her slave, Thomas Shaw. Shaw enlisted in the 65th Colored Infantry on January 19, 1864 and served until 1866. While serving with the Ninth Cavalry in Indian Territory on August 12, 1881, Shaw showed unusual heroism in action at Carrizal Canyon, New Mexico. His citation stated that he forced the enemy back after stubbornly holding his ground in an extremely exposed position, and prevented the enemy's superior numbers from surrounding his command.

Company K, Ninth Cavalry Regiment under the leadership of Captain Parker had made contact with Apache leader, Nana and his Indian warriors. The Company was outnumbered by the large force of attacking Indians. Sergeant Shaw was serving as acting first sergeant. Shaw demonstrated outstanding leadership and courage when he forced the enemy to retreat and prevented the warriors from encircling his company. Later, Captain Parker would write that Sergeant Shaw "provided me support and was near me to give assistance if I would fall". The superb leadership abilities and gallant action of Shaw portrayed the outstanding qualities of those "Real Buffalo Soldiers of yesteryear who assisted greatly in the conquest of the Western Frontier with the "white majority Americans".

Sergeant Thomas Shaw served in the United States Army for almost 30 years. He retired 100 years ago, in 1894 at Fort Myer, Virginia and lived in Rosslyn, Virginia during his retirement. He died in 1895 and is buried in Arlington National Cemetery.

The life of Sergeant Thomas Shaw is quite interesting when one examines his exploits in the military and his desire to achieve and accomplish his goals with superior results. We can also examine a very unique quality that Shaw possessed and that was his belief in family and the extended black family. He also believed in the pursuit of excellence through education. I make this statement because somehow in his leadership qualities, he must have transferred them to his family life. Because his grand daughter became a very successful school teacher in the District of Columbia Public Schools and his grandson must have inherited those unusual qualities also. Because today Sergeant Thomas Shaw's great grandson has carried the torch of excellence, leadership and the pursuit of excellence in education. Shaw's grandson is Colonel (Doctor) John W. Hughley III, a graduate of Howard University's College of Medicine. Hughley is a member of the Army's U.S. Medical Corps and the state surgeon for the District of Columbia army's National Guard. He is an outstanding military surgeon and a successfully civilian physician. He is the former chief, outpatient medical clinic, Walter Reed Army Medical Center, Washington, D.C. Yes, his great grandfather was definitely a hero and a family patriarch. I nominate Sergeant Thomas Shaw for an image model for all young people today of all races to emulate and understand that faith, belief in one's self, courage and the personal desire to overcome obstacles can eventually lead a person to obtain their personal successes and endeavors. A former slave, a soldier, a father and a great grandfather named Thomas Shaw did it so eloquently.

James Shears
Sergeant, Company E, Tenth U.S. Cavalry Regiment

James Shears was a member of the Tenth U.S. Cavalry. He was from the state of Georgia. Shears died March 17, 1922. He was buried in the Soldiers' Home National Cemetery, Washington, D.C.

Edward Simms
Private, Ninth U.S. Cavalry Regiment and Twenty-fifth Infantry Regiment

Edward Simms was born in Buton, Canada. He was the son of Isaac Simms and Maria Touer Simms. His mother died in the 1870's. Edward had a brother, William Simms. His father, Isaac was born a slave in the state of Maryland and escaped from slavery prior to the 1860's by the Underground Railroad route to Canada. His family later moved to Oberlin, Ohio.

Private Edward Simms enlisted in the Ninth U.S. Cavalry in 1887 and served on the Western frontier. In 1895, Simms enlisted in the Twenty-fifth Infantry Regiment and served during the Spanish American War in 1898. After the Spanish American War, Edward Simms was discharged and returned to Oberlin, Ohio. In 1900, Simms moved to Akron, Ohio where he lived until his death in 1938.

John Simms
Private, Tenth U.S. Cavalry Regiment

John Simms was a member of the Tenth U.S. Cavalry Regiment. Simms died on September 11, 1920. He is buried in Arlington National Cemetery.

Columbus Smith
Private, Company L, Twenty-fifth Infantry Regiment

Columbus Smith was a member of the Twenty-fifth Infantry. Smith served in the Spanish American War. He died January 1, 1970 and is buried in the Soldiers Home National Cemetery, Washington, D.C.

George H. Smith
Cook, Troop F, Tenth U.S. Cavalry

George H. Smith was born in Virginia on July 4, 1862. Smith served during the Spanish American War. He died on September 2, 1941 and is buried in Arlington National Cemetery.

Richard Smith
Private, Company E, Twenty-fifth Infantry Regiment

Richard Smith was born in 1860 at Pulaski, Virginia. He was Twenty-six years when he was enlisted in the Army on December 7, 1866 at Cincinnati, Ohio, by Lieutenant O'Connell. The official records listed his description as "black eyes, and hair and black complexion". Richard Smith was discharged at the expiration of his term of service on December 6, 1891, at Fort Buford, North Dakota.

Howard Snowden
Private, Tenth U.S. Cavalry Regiment

Howard Snowden was born in Howard County, Maryland. He was enlisted in the Army at the age of twenty-three, on December 21, 1886 at Baltimore, Maryland, by a Captain Overton. He civilian occupation was a laborer. The official records listed his description as "eyes, hair and complexion, black". Snowden was discharged on March 31, 1891 at Fort Apache because of disability.

Robert Spankler
Private, Company H, Twenty-fourth Infantry Regiment

Robert Spankler enlisted in the Twenty-fourth Infantry Regiment. Spankler served during the Spanish American War. He is buried in Arlington National Cemetery.

Percy D. Spence
Private, Tenth U.S. Cavalry

Percy D. Spence enlisted in the Tenth Cavalry Regiment. He was born on April 12, 1871. He was from the state of Pennsylvania. Spence died on March 9, 1951 at the age of 80. He "Rests Among the Known" at Arlington National Cemetery.

Emanuel Stance
First Sergeant, Company F, Ninth U.S. Cavalry Regiment

Emanuel Stance was born in Carroll Parish, Lousiana in 1847. He enlisted in the Ninth Cavalry in 1866 at the age of nineteen. Stance earned a Medal of Honor during the Indian Campaigns at Kickapoo Springs, Texas on May 20, 1877. He was given the responsibility to command a detachment of eight Buffalo Soldiers. They were pursuing a band of Indians who had captured or taken two children of a white settler, Philip Buckmeier of Loyal Valley, Texas. Stance and his men had left Fort McKavett, Texas on May 19, 1870 and they had traveled some 14 miles before they spotted some Indians moving across the hills with horses. They were able to capture some horses as the Indian band fled. On their way to Fort McKavett, the detachment sighted about 20 Indians racing toward two teams of government horses with guards. Stance and his men fired some shots and the Indians decided to confront the detachment.

Emanuel Stance and his nine men engaged in a brief skirmish with the Indians near an area called "Eight Mile". Stance's performance was considered excellent by the post commander, Captain Henry Carroll. Sergeant Emanuel Stance had been promoted to First Sergeant at the time of his death. This courageous Buffalo soldier was murdered on December 25, 1887 near Fort Robinson, Nebraska. He was the first known black soldier to received the Medal of Honor while fighting in the Far Western frontier.

Jacob W. Stevens
First Sergeant, Twenty-fourth Infantry Regiment

Jacob W. Stevens enlisted in the Twenty-fourth Infantry Regiment. He was born on September 10, 1864. Stevens served during the Spanish American War. He was married to Julia Stevens who was born on September 23, 1883. Jacob W. Stevens died at the age of 65 years on March 29, 1929. Jacob and Julia Stevens are buried in Arlington National Cemetery.

John Sumner
First Sergeant, Ninth Tenth U.S. Cavalry Regiment

John Sumner was a member of Troop D, Ninth U.S. Cavalry Regiment. He was born on October 15, 1868. Sumner was married to Theresa L. Sumner who was born on October 29, 1874. John Sumner died at the age of 57 on February 24, 1931. His wife Theresa died at the age of 77 on December 10, 1951. John and Theresa Sumner are buried in Arlington National Cemetery.

George H. Tancil
Sergeant, Troop F, Ninth U.S. Cavalry Regiment

George H. Tancil was a member of the Ninth Cavalry Regiment. He was from Washington, D.C. Tancil was born on August 8, 1874. He was married to Mary Etta Tancil who was born on July 18, 1874. She died at the age of 84 years on May 9, 1958. George and Mary Etta Tancil are buried in Arlington National Cemetery.

Harvey A. Thomas
Trooper, Ninth U.S. Cavalry Regiment

Harvey Thomas was born on July 24, 1863 in Columbus, Ohio. He attended public schools in Ohio and matriculated at Fisk university and Lemoyne College (known later as Lemoyne-Owen College), Memphis, Tennessee. In 1883, he enlisted in the Ninth U.S. Cavalry and served until 1888. After his discharge, the former Buffalo Soldier studied at Meharry Medical College in Nashville, Tennessee for two years before moving to Chicago, Illinois. Thomas joined the Eight Illinois Volunteer Infantry Regiment and served as the regimental Adjutant during the Spanish American War period.

William M. Thomas
Private, Troop E, Tenth U.S. Cavalry Regiment

William M. Thomas was a member of the Tenth Cavalry Regiment. Thomas was born on April 14, 1876 and he died at the age of 65 years. on February 24, 1941. Thomas is buried in Arlington National Cemetery.

George Bennett Thornton
Trooper, Tenth U.S. Cavalry Regiment

George Bennett Thornton was born in Aberdeen, Ohio on March 1, 1881. He was the son of Thomas and Celia Thornton. Thornton married Lydia Davis of Columbus, Georgia. George Thornton studied at Walden University, Nashville, Tennessee and Tuskegee Institute, Alabama. Thornton enlisted in the Tenth U.S. Cavalry Regiment and served as a Trooper and

Bandsman. He served in the military during the period 1900-1907. After his discharge he pursued a career in music. Thornton was a band and orchestra teacher at the Agricultural and Mechanical School, Pine Bluff, Arkansas, Southern University, Baton Rouge, Louisiana and Wilberforce University, Wilberforce, Ohio. George Bennett Thornton was a former Buffalo soldier who attained excellent achievements in his civilian career of music education.

William Toliver
Private, Tenth U.S. Cavalry Regiment

William Toliver enlisted in the Tenth Cavalry Regiment. He was from the state of Pennsylvania. Toliver died on February 24, 1927. He responded to the "call of taps" and is buried in Arlington National Cemetery.

Charles Burrill Turner
Sergeant Major, Tenth U.S. Cavalry Regiment

Charles Turner was born on June 25, 1859 at Mineral Point, Wisconsin. He received his formal education in Cincinnati, Ohio. On November 15, 1875 at sixteen years of age, he enlisted in the regular army at Indianapolis, Indiana. He was assigned to Troop E, Tenth U.S. Cavalry Regiment, in May, 1876, as a troop clerk where he joined the unit at Pecos River, Texas. Sergeant Major Turner was with the Tenth Cavalry in operations against the Indians from September 1876 to June, 1879 and he served on the Mexican border at San Felipe.

In 1880, Turner was present at the disarming of the Mescalaro Apache Indians in New Mexico and in the same year he participated in the campaign against Victoria's band of Indians. Turner's combat campaigns also included operations against the Kiowas and Comanches in the Indian Territory during July 1881. He was with the regiment when it participated in chasing and capturing Geronimo. In January, 1888, Turner passed a board of officers for the position of Commissary Sergeant. He was a delegate to the National Regular Army and Navy Union Conventions of St. Louis, Missouri in 1893. During the Spanish American War, Turner was on regimental recruiting duty in the state of Kentucky. On November 6, 1898, Turner was appointed a Sergeant Major.

Charles Tyler
Corporal, Tenth U.S. Cavalry Regiment

Charles Tyler enlisted in the Tenth U.S. Cavalry Regiment. He attained the rank of Corporal. Charles Tyler a Buffalo Soldier of the Tenth Cavalry died and is buried in Arlington National Cemetery where he "Rest Among The Known". He was from the state of Virginia.

James E. Tynes
Trooper, Tenth U.S. Cavalry Regiment

James E. Tynes was born on May 22, 1872 in Smithfield, Virginia. He enlisted in the army on May 25, 1896. Tynes served during the Spanish American War. When he arrived at Chickamauga, he was detailed as mail carrier. He also was the post school teacher. He was left by his regiment at Port Tampa and had to report aboard the Flagship Seguranca with General Shafter and his staff. They landed in Cuba three days before Tynes' regiment. It is believed that James E. Tynes was the first black soldier to land in Cuba. He thought his regiment had landed at another point, and marched 18 miles through Cuba alone and without his weapon.

William A. Vrooman
Quartermaster Sergeant, Ninth U.S. Cavalry Regiment

William Vrooman enlisted in the army on June 18, 1886 at Buffalo, New York and was assigned to Troop I. Ninth U.S. Cavalry. He was promoted to sergeant on January 22, 1887. Vrooman served in the Sioux Indian Campaign in September, 1888, and participated in the engagement at Drexel Catholic Mission and Wounded Knee, from December 1890 to January 1891. Vrooman was a superior marksman and was a competitor in the Tri-Department Rifle and Pistol competition in 1892, winning second, third, and fifth department rifle medals, and seventh department pistol medals. He served in Cuba during the Spanish American War and in the Philippines from 1900-1902 and 1907-1908. William Vrooman had over twenty-four years service with the army and in January 1909 had attained the rank of regimental quartermaster sergeant.

James Waites
Private, Tenth U.S. Cavalry Regiment

James Waites enlisted in the Tenth U.S. Cavalry Regiment. He died on November 25, 1927 and is buried in Arlington National Cemetery.

Augustus Walley
Private, Company I, Ninth U.S. Cavalry Regiment

Augustus Walley was born in Reistertown, Maryland in 1856. He enlisted in the Ninth Cavalry Regiment at the age of 22 years in 1878. Walley displayed bravery in action with hostile Apache Indians on August 16, 1881, at Cuchillo Negro Mountain, New Mexico. He was awarded the Medal of Honor on October 1, 1890. When one soldier was trapped and could not move from his position during a fight with the Apaches, the lieutenant of Company I, asked Private Walley to go to the rescue of the soldier. Walley mounted his horse and rode under fire to the stranded man. He dismounted, helped the soldier in the saddle, then mounted behind him and rode toward the rear to safety. Walley also served during the Spanish American War and was a member of the Tenth U.S. Cavalry. He was recommended the second time for a Medal of Honor for assisting in the rescue of a wounded officer under heavy enemy fire. The recommendation was denied. Augustus Walley served in the Philippines in the 1900's. Walley retired as a First Sergeant in 1907 and lived in Butte, Montana. Augustus Walley was a "Real Buffalo Soldier" who served in the Indian Wars, Spanish American War and the Philippine Insurrection and retired after almost 30 years of a commendable military career.

George H. Wanton
Private, Troop M, Tenth U.S. Cavalry Regiment

George Wanton was born in Paterson, New Jersey and entered the service from there. During the Spanish American War, he was stationed at Toyabacoa, Cuba, and on June 30, 1889, exhibited unusual valor, for which he was awarded the Medal of Honor on June 23, 1899. His citation stated that he voluntarily went ashore in the face of the enemy and aided in the rescue of his wounded comrades, this after several previous attempts at rescue had been frustrated.

John Ward
Sergeant, Detachment of Seminole Negro Indian Scouts, U.S. Army

John Ward was born in 1847 at Santa Rosa, Mexico. He enlisted in a detachment of Seminole Negro Indian scouts on August 16, 1870 and served during the period 1870-1894, receiving his final discharge on October 5, 1875, as a sergeant, at Fort Ringold. According to his official description he was five feet seven inches in height and had black hair, eyes and complexion.

On April 25, 1875, he displayed unusual gallantry in action with the Indians. He was with Lieutenant John L. Bullis of the Twenty-fourth Infantry Regiment, Isaac Payne, and Trooper Pompey Factor when they struck a trail of about seventy-five horses and followed it to the Eagle's Nest crossing of the Pecos, coming upon the Indians as they were attempting to cross to the western side with the stolen herd. The party tethered their mounts and crept to within seventy-five yards before they opened fire, which they maintained for about forty-five minutes, killing three warriors, wounding a fourth, and twice forcing the raiders to retire from the herd. As the scouts reached their horses, mounted, and were preparing to leave, Sergeant Ward noticed that Lieutenant Bullis' mount, a wild and badly trained animal, had broken loose, leaving him without a horse among the Indians who were rapidly approaching. Sergeant Ward dashed back to the lieutenant's aid followed by his comrades. The Indians opened fire on the rescue party and a bullet cut Ward's carbine sling. As he was helping the lieutenant to mount behind him, a ball shattered the stock. Factor and Payne had meanwhile been fighting off the Indians, and the three scouts and the lieutenant were able to make their escape, saved, as Lieutenant Bullis wrote, *"by a hair."* John Ward received the Medal of Honor for his bravery and devotion "beyond the call of duty" in this encounter. John Ward died on March 24, 1911, in the Seminole Camp on a government reservation, and was buried on March 26, 1911, in the Seminole Negro Cemetery on Las Moras Creek, Kinney County. He was nearly sixty-four years old at the time of his death. He and his wife, Judy, had four children.

William Chapman Warmsley
Surgeon, Tenth U.S. Cavalry Regiment

William Warmsley was born on October 20, 1869, in Gloucester County, Virginia. In 1876, his father moved to Norwich, where William attended public schools and the free academy. On March 12, 1887, he enlisted in the Ninth U.S. Cavalry, and was assigned to Troop D. He served five years during which he was post teacher and post sergeant major. He attended sessions at the College of Medicine, Howard University, from which he graduated in 1898.

In the Spanish-American War, Warmsley served as a surgeon with the Twenty-Third Kansas Volunteers, acting assistant surgeon with the Ninth Louisiana Infantry, Fifth U.S. Infantry and Tenth U.S. Cavalry regiments. On September 9, 1899, he was appointed assistant surgeon in the Forty-Ninth U.S. Volunteers. He was confronted with problems of yellow fever in Cuba and dysentery and malaria in the Philippines. On April 24, 1900, he was stationed at Luzon in the Philippine Islands.

Bob Washington
Sergeant, Company C Twenty-fourth Infantry Regiment

Bob Washington enlisted in the Twenty Fourth Infantry Regiment. He was born on June 18, 1884 and was a native of Virginia. Washington was married to Elizabeth Washington. Bob Washington died on August 4, 1965. His wife Elizabeth died on October 26, 1944. Bob and Elizabeth Washington are buried in Arlington National Cemetery.

John T. Wiley
Private, Troop H, Ninth U.S. Cavalry Regiment

John T. Wiley enlisted in the Ninth Cavalry Regiment. He also served during World War I. Wiley was married to Maude Wiley who was born on July 2, 1898 and died on June 28, 1986 at the age of 88 years. John Wiley died July 11, 1940. He and his wife Maude are buried in Arlington National Cemetery.

Benjamin West
Private, Tenth U.S. Cavalry Regiment

Benjamin West enlisted in the Tenth U.S. Cavalry Regiment. He was married to Martha West who was born on March 25, 1880. She died on June 17, 1953. Private Benjamin West, a Buffalo Soldier who was assigned to Troop F, Tenth Cavalry Regiment and his wife Martha are buried in Arlington National Cemetery.

Arthur Williams
Sergeant, Ninth U.S. Cavalry

Arthur Williams was born in 1870 at Athens, Georgia. He attended public schools until he was eighteen and later moved to Atlanta, Georgia. He enlisted in the Ninth U.S. Cavalry and was stationed at Jefferson Barracks. He was promoted to sergeant and served as a drill master for eight years. In the 1890's, he enlisted in the volunteer regiment and was commissioned a lieutenant.

George Washington Williams
Sergeant Major Company L, Tenth U.S. Cavalry Regiment

George Washington Williams was born on October 16, 1849 in Bedford Springs, Pennsylvania. He was the son of Ellen Rouse Williams who was of black and German parentage and Thomas Williams who was of black and Welsh descent. Williams received his elementary education in Pennsylvania and attended secondary school in Massachusetts. After four years at Newton Center, a School of Technology, he was enlisted in the Union Army by Major George L. Sterns. Williams ran away from home at the age of fourteen and using the name of a relative, he was accepted for enlistment. Williams served during the Civil War in Company C, 41st United States Colored (USC) Infantry. He was promoted rapidly from private to Sergeant Major. He was severely wounded in the assault on Fort Harrison and discharged from the service, but reenlisted and was detailed to the staff of General Jackson in 1863 and accompanied him to Texas. Williams was mustered out of the service in 1865. He then immediately enlisted in the Mexican army and was given the rank of orderly sergeant, First Battery, State of Tampico.

A week later, he was appointed to the rank of lieutenant colonel and given the position of assistant inspector general, artillery. After the capture of Maximilan in 1867, Williams returned to the United States. On August 29, 1867, he enlisted in the regular U.S. Army and was assigned to Company L, Tenth U.S. Cavalry Regiment, Fort Riley, Kansas. He was promoted to drill sergeant and later sergeant major. Williams was assigned in Indian Territory at Fort Arbuckle. On September 4, 1868, Williams was discharged from the army. He had received a gunshot wound through his left lung in the line of duty. Being unfit to perform his duties, he received an honorable discharge.

George Washington Williams had a very brief military career at a young age. However, he enjoyed a very interesting and productive civilian career. He was a successful minister, editor and publisher and world traveler. He was pastor of the Twelfth Street Baptist church in Boston, Massachusetts. He also served briefly as a state representative in the Ohio state legislature. Williams had traveled to Africa, Belgium and England. He was the author of *The History of the Negro Race in America from 1819 to 1880, Negroes as Slaves, as Soldiers and as Citizens and An Historical Sketch of Africa and an Account of the Negro Government of Sierra Leone, Africa.* Williams is buried in Layton Cemetery, London, England.

Isaiah Williams
Private, Twenty-fifth U.S Infantry Regiment

Isaiah Williams enlisted in the Twenty-fifth U.S. Infantry Regiment. Williams died November 12, 1927. He is buried in the Soldiers' Home National Cemetery, Washington, D.C.

Joseph Williams
Private, Tenth U.S. Cavalry Regiment

Joseph Williams enlisted in the Tenth U.S. Cavalry. Williams died September 26, 1892. He responded to the "call of taps" and is buried in the Soldiers' Home National Cemetery, Washington, D.C.

Moses Williams
First Sergeant, Company I, Ninth U.S. Cavalry regiment

Moses Williams was born in Carrollton, Lousiana in 1845. He enlisted in Company I, Ninth U.S. Cavalry Regiment at the age of 21 years in 1866. His civilian occupation was a farmer. Williams was awarded the Medal of Honor on November 12, 1896 for heroically rallying his detachment at the foothills of the Cuchillo Negro Mountains, New Mexico on August 16, 1881. His citation read : *"Rallied a detachment skillfully and conducted a running fight of three or four horses and unflinching devotion to duty in standing by his commanding officer in an exposed position under a heavy fire from a large party of Indians, he saved the lives of at least three of his comrades."*

Williams retired in 1898 after completing 32 years of military service. He lived in Vancouver, Washington after his retirement. Moses Williams died in 1899 and was buried in Vancouver Barracks Cemetery.

William H. Williams
Private Ninth U.S. Cavalry Regiment

William H. Williams was a native of Virginia. He was a member of the Ninth U.S. Cavalry Regiment. Williams died on April 27, 1940 and "Rest Among The Known" in Arlington National Cemetery.

James E. Wilson
Private, Tenth U.S. Cavalry Regiment

James E. Wilson was a member of the Tenth U.S. Cavalry. Wilson was from the state of Massachusetts. Wilson died July 1924. He is buried in the Soldiers' Home National Cemetery, Washington, D.C.

Willis Wilson
Private, Company C, Twenty-fourth Infantry Regiment

Willis Wilson was a member of the Twenty-fourth Infantry Regiment. He was from the state of Louisiana. Wilson served during World War I. Willis Wilson was born on September 15, 1881 and died on March 27, 1932. He is buried in Arlington National Cemetery.

William O. Wilson
Corporal Company I, Ninth U.S. Cavalry Regiment

William O. Wilson was born in 1868 at Hagerstown, Maryland. He enlisted in the Ninth Cavalry Regiment at St. Paul, Minnesota on August 21, 1889. His civilian occupation was an upholsterer. In the 1890's some Indians living at the Rosebud, Cheyenne River Standing Rock, and Pine Ridge Reservations had adopted the "Ghost Dance Religion", which was becoming popular among the Sioux Indians, especially in the Dakotas. There were some Indians at the Pine Ridge Reservation in November 1890 who were excited during their dance ceremony and an Indian agent feared a possible rebellion by the Sioux Indians. Major General Nelson Miles decided to order many units of his command to the areas of the reservation to contain an Indian leader and his followers. The eventual disarming of Big Band and his Indian warriors led to the massacre of many Indians at Wounded Knee in December 1890.

The Ninth Cavalry's companies F,I and K were sent to the Pine River Reservation to provide assistance to the Indian agent in November 1890. They were under the command of Major Guy V. Henry. Henry also had a white unit under his command, Troop D of the Twenty-sixth Cavalry. (This was interesting because a white Troop was riding and prepared to fight along with the black troops of the Ninth Cavalry in 1890, and because of the U.S. Government extreme acts of racism in 1918, black military soldiers would have to fight beside the French Troops and their colonial soldiers the Senegalese). Major Henry had traveled ahead of his wagon train which was left with the soldiers of Company I and Captain S. Lord was attacked by some Indian warriors and it was necessary for Captain Henry Lord to dispatch a courier immediately to obtain assistance from Major Henry.

Corporal William O. Wilson volunteered for the daring mission which he successfully completed. His gallant performance possibly saved the lives of many members of Company I and the wagon train soldiers. In March 1891, Wilson was found guilty of being absence without leave and theft of a rifle. He was confined briefly and after writing a letter to the president, it appears that the commanding general of the Military Department of the Platte ruled that the sentence be disapproved and Wilson be returned to duty. Unfortunately, while Wilson was in Denver, Colorado on September 5, 1893, participating in a rifle competition, he made a decision to desert from the army.

Wilson was a Real Buffalo Soldier who performed courageously on the battle field. However, he faced some serious personal problems and they were evident in this soldiers decision to desert from the army, his proud unit and possibly many good friends.

Henry C. Winston
Private, Twenty-fifth Infantry Regiment

Henry C. Winston was a native of Virginia. Winston died December 26, 1936 and is buried in Arlington National Cemetery.

Brent Woods
Sergeant, Company B, Ninth U.S. Cavalry Regiment

Brent Woods was born in Pulaski, Kentucky in 1850. He enlisted at the age of 23 years in 1873 in Louisville, Kentucky. His civilian occupation was a farmer. Woods was awarded the Medal of Honor on July 12, 1894 for outstanding bravery in assuming command of his detachment when the commander, Lieutenant George W. Smith was killed by the attacking Indians. Brent Woods ordered the soldiers to dismount and move toward a nearby hill where they were able to continue firing toward the Apache Indians. Under the leadership of Sergeant Woods the detachment was able to cause the Indians to retreat. Woods prompt actions, correct decision and excellent leadership saved the lives of his men and some civilian miners who were in the immediate area where the fighting occurred. Brent Woods retired from the army in 1902. He served a total of 25 years. The outstanding NCO with superb qualities of leadership died in Somerset, Kentucky in 1906.

James A. Wright
Private, Company H, Twenty-fourth Infantry Regiment

James A. Wright was born on December 30, 1895. He enlisted in the Twenty-fourth Infantry Regiment. He was married to Ruth S. Wright. Private Wright served in the Military during World War I. He died on November 5, 1927. Ruth S. Wright died on December 27, 1968 at the age of 68 years. James and Ruth Wright are buried in Arlington National Cemetery.

CHAPTER 12

SAGA OF MARK MATTHEWS

THE SAGA OF MARK MATTHEWS
FOREWORD

I had the distinguished pleasure of meeting and interviewing a "Real Buffalo Soldier", First Sergeant Mark Matthews who is currently residing in Washington D.C. I am also grateful to his daughter, Mrs. Mary Matthews Watson for assisting me in making the visits and interviews possible. The experience was historical, captivating, and most informative. It was amazing to listen to Mark Matthews recall in detail experiences of his early childhood and military experiences. His vast knowledge of geographical details and events that occurred some seventy years ago is commendable. I was quite surprised during an interview when Mark Matthews was explaining the position he would move to when firing his weapon and assuring himself that he would directly hit the center of the target. Matthews immediately assumed the position on the floor and then returned to his seat. His reflexes and swiftness was an experience to witness.

First Sergeant Mark Matthews had a very challenging and productive military career. He was able to find sufficient time for his family. Matthews was married to his late wife, Genevieve Pearl Hill for 52 years. They are the parents of Mary, Mark Jr., Gloria Joyce, Shirely Ann and Barbara Jeannette. Mark Matthews also has an extended family of grandchildren and great grandchildren. He had several family members who served in the U.S. military services, possibly following a tradition of a "Real Buffalo Soldier". Mark Matthews has been blessed to enjoy many fruitful years of life. On August 7, 1994, he celebrated his birthday and the attainment of 100 years. Since I was able to hear the stories of First Sergeant Mark Matthews' life as told by him, I have decided to present in the first person, his story, the story or saga of this outstanding veteran of the Buffalo Soldier era, to you, the readers. Mark Matthews gave his best as a soldier, his civilian life has been very successful and his military career, a historical rarity. Because for a man who reached the age of 100 years this year and was able to relate this story as accurately clearly and interesting is most unusual. I say to the readers of this manuscript, please read on and on the true story of a man who actually lived these monumental experiences during the period 1894-1994.

Robert Ewell Greene
August 4, 1994

I was born on August 7, 1894 in Greenville, Alabama. My mother died when I was a young baby. My twin sister died from dropsy at three years of age. When I was 5 years old, several aunts wanted me to live with them. There was one aunt who was attending school and was married to a school teacher. They would frequently move to other towns. My father had married a second time, and my mother was his second wife. I had two half brothers from my father's first marriage. All of my aunts wanted me to live with them. They always would say "Let me carry Mark home. I would like to keep Mark". I finally ended up with an aunt who lived in Louisville, Kentucky. When she became ill and died, I was suppose to live with an aunt in Baltimore, Maryland; however, I never went to Baltimore to live with her.

It was very different for me to move from one place to another. When I was 13 years of age. I decided to go to Lexington, Kentucky.

As a young boy, I was very fond of ponies and horses. I had a little pony and I would deliver my newspaper on my pony. The name of the newspaper was *Saturday Evening Blade*.

In 1913, while on a school vacation, we went to Lousiana, a place where the Mississippi river enters into the Gulf of Mexico. The place was called Dunbar, Lousiana, near Bay St Louis, Mississippi. While visiting in Lousiana, there was a big storm that occurred between Bay St. Louis, Mississippi and Baton Rouge, Lousiana. The highways and railroads were flooded with water. The Lousiana and Nashville railroad tracks had to be repaired before we were able to leave the town. We were there for one week.

There were several houses on a large hill where you could see the front of the house and could not observe the back area. When you would go toward the back, you could see that the house was built above the water. Because the people could look directly out their back door and see only water below, the Pearl river. Later, we went to Louisville, Kentucky. As a child in Louisville, I would attend a class and I loved to go to a place called Corney. There was a memorial to Daniel Boone, the great Indian fighter. There was a large tree that stood at this place, but it was cut down to its stump. They had a plague and picture explaining about Daniel Boone.

Corney was near a railroad track that went to Lexington, Kentucky. But if you crossed the mountain you would be in West Virginia. There was no road, you could walk or travel on a burrow.

When I was living in Lexington, Kentucky, I would go down to the race track. I loved horses. I could ride almost any horse. I was comfortable with them because I was raised up with them. The man in charge of the horses at the race track was Jack Terrell. He gave me a job where I would exercise the ponies and horses. Sometimes I would ride a single footer. You see a single footer is a horse that when it is running, he will watch each step as he moves. Now, a pacer will not do that. He is a gallop horse. I trained a pacer that was also fast. If you wanted to stop him, just talk to him. I could handle him so nice. The man in charge would always pat me on the shoulder and say I was riding very good with the horses. He would also take a strap and tie it around the horse and around my knees. Because the horse could run so fast that I could have fallen if I did not have the straps tied around me. I would train and exercise the horses to be ready for the jockey. Yes, they had a few black jockeys. There was one black jockey that was so good, that some English people saw him perform and took him to England and he never came back to the race track.

Some soldiers came to town around 1913 and they were coming from Fort Ethan Allen, Vermont and said they were in the Tenth Cavalry Regiment. It has been said that when President Theodore Roosevelt became president, he gave the Tenth Cavalry Regiment a trip abroad. When they returned they were stationed at Fort Ethan Allen. Many people had never heard of these soldiers. I met some of them when they came through going toward the Mexican border to pursue Poncho Villa and his men. When they came down to the race track, they asked me about the best horses they should bet on. I told them that I could not give any information about the horses because I would be doing wrong. I said I cannot tell you which one is the fastest. As we walked pass the horse stables and they were looking at the horses, I would say this is a good horse and pat another horse and say the same, but I would never name the horse or say this is the fastest.

When the military started calling for soldiers to enlist in the army. I went to see my boss. I told him that I wanted to go where those soldiers were going. The soldiers had told me, that they ride the horses when they are feeling good.

Later, I told my supervisor that if I would go home, my Daddy would not let me leave again. He said, let me handle it. He knew a man who recruited soldiers and he was his friend. He said my friend will take your word for anything you tell him. Jack Terrell wrote a letter for me to take to this man at the recruiting station in Huntington, West Virginia, 15 miles from Lexington, Kentucky. The recruiting station was located on the second floor above the First National Bank. I gave the man the letter. He said take a seat. There he began to examine me. I was told to take off my clothes and to pick up one leg and hop towards him. Then I had to turn around and walk straight towards him. Then I walked down the hallway and back. I was given a big book and told to read the first few lines. I started reading the constitution and bylaws of the United States "I will obey and protect.." The recruiting man said stop, you are okay and will do fine, please stop reading. I was then given a card to fill out and write down my father's name and some other information. I was directed to go down stairs and use the restaurant for my meals and I was shown where I would sleep that night. The recruiting man said the next morning I would board a train for Columbus, Ohio where I would be sworn in to the U.S. Army. The next morning as I boarded the train, I saw three men who were also going to Columbus to join the army.

Upon my arrival at the camp in Columbus, Ohio, I was issued my military uniform, assigned to quarters and given several lectures on the rules and regulations of the army. We were sworn in by an officer.

The following day we had to fall in formation and run double time for a distance. Later, we were given a book to read on the general rules. We had to know all of the rules before we could go outside and move around the camp. Finally, I was shipped out to Fort Huachuca, Arizona. On the way to Arizona we stopped at Columbus, Mexico, where some soldiers were leaving Mexico. I had to join these troops and then we proceeded for Fort Huachuca. I reported to a recruit camp where the young recruits were trained. We had to study the 1916 Drill Regulations book. You had to learn the regulation book, everything in it. When you finished, there was nothing they had to tell you. We were assigned to a troop. My first troop was the machine gun troop, Troop M. On the 4th of July the men would have a nice dinner with their troop. During my early training, the noncommissioned officers (NCO) would always be watching you, to be sure you were doing everything okay.

I was introduced to my first horse. All of the horses were in a large circle in the corral. We were given instruction about the horse, learned about their physical body and we had to observe them. We were explained how to use the morning and sick reports for the horses. I felt quite confident about horses because I knew something about them. My first horse was named Malachi. The first letter of the horse's name would always be the letter of the troop that you were assigned.

While serving in the cavalry, I became a very good marksman. I worked hard to become an expert. I was the best marksman in my troop for nine years. I competed in pistol and rifle competitions and always won first place. I have even shot against the Police and Rifle teams of Los Angeles, California and Rifle teams of Arizona and Fort Sam Houston, Texas. You can find out who won in a particular competition over the years, by visiting a records center in Kansas City that maintains all these records. They have records since the competition started, they go back before you were born and before I was born. They have records of every event.

When World War I started, I was still stationed at Fort Huachuca, Arizona. There were 414 soldiers in my recruit camp. When the armistice was signed, only seven men were still stationed at Fort Huachuca. The other men were shipped to Illinois and other places and eventually went to France. Those men no longer had their horses. I still had my horse. Every year the military would send men to Fort Hauchuca and assign them temporarily to the Tenth Cavalry Regiment. Later, they would be transferred to go overseas to France. The first sergeant would order the men to fall in formation. He would have them count 1 to 4, then he would order all the men who were number one to step one pace to the front then he would continue this procedure for two, three and four at different times. The men with the particular numbers he called, were told to report to the adjutant. The adjutant would issue them orders to be shipped to France. I was never called and remained as a standing cadre. I would always have a different number from those numbers called. However, one day the first sergeant called my name along with 17 other soldiers. We reported to the adjutant and were issued orders to report to Deming, New Mexico. I was a "one handed barber" and was assigned to the Post Barber Shop, located near McCordy Hospital. I worked in the Barber Shop until the armistice was signed.

The Ninth Cavalry left Douglas, Arizona in 1917 and went to the Philippines. They stayed there until 1921 and returned to Forts Huachuca and Riley. Some of them had their wives and children at Fort Huachuca.

After the Armistice was signed, the army said that soldiers could enlist for 7 years, do seven years, or be discharged. They also could reenlist for 3 years and go to the Philippines or reenlist for 14 years and get a nine dollar bonus. I left for New Mexico to serve on the border with the immigration officer at Naco, Arizona. Later, I would do patrol duty.

Sometimes there would be little uprisings on the Nogales side in Mexico. There were two men trying to be mayor of the towns in Mexico. They had a gold mine at Cananea. The gold would be taken to Bisbee to be cooked and prepared for solid gold. This was around 1921. The two men, Escobel and Escapia wanted to be mayor of the Sonora state. The Governor of Arizona wanted protection for residents in the area because there was fighting on the Mexican side. Fort Huachuca was 68 miles from Nogales. It was in 1919 when the Tenth Cavalry was involved the last time in those uprisings on the Mexican side. Governor Churchill of Arizona had received complaints from residents on the American side. They claimed that bullets were coming on their side from Mexico and going through their homes. They wanted protection. One night while we were in the theatre, an announcement was made. All men of the machine gun troop report at once to your orderly room. When we arrived at the orderly room, the first sergeant was standing there and had a pad in his hand. He said "Saddle up". We immediately obtained our rifles. We arrived at Nogales and made a show of force. Orders were issued to the Mexicans to move 5 miles back into Mexico.

After World War I, I was assigned to an immigration office along the border. For a long time all I had to do was to get up in the morning, eat my breakfast, shine my boots and then report to the immigration officer. I did duty with the immigration officer for a pretty good while. When a Mexican wanted to enter the United States, especially if he was running from some trouble he had in Mexico, all he had to do was show his passport. The passport had the necessary information, his picture and the right for him to enter and leave U.S. However, the immigration officer could not stop him. The military could stop him. In those days the immigration officer did not have that authority. As a military man on duty, I could stop the Mexicans and have them returned to Mexico. The white cavalry units were stationed in El Paso, Texas and were performing the same duties we were.

Around 1920, some information was posted about vocational training for all soldiers. My first sergeant asked me if I was interested in attending a training school. I said yes, and received orders to attend Saddlers' School at Fort Sam Houston, Texas. I was taught how to make things out of leather. I can make anything with leather, all I need is the leather and the tools. I have a riding crock that I made and you would never believe that I made it.

I completed the school and returned to my unit. A few months later, my first sergeant asked me if I would like to attend an advanced saddler course at Fort Riley, Kansas. Again, I told him yes, I reported to Fort Riley and was assigned to headquarters troop for training purposes. I met the director of the school, Brigadier General Churchill. I stayed there several months. They gave me an examination on the different subjects we studied and you had to explain the things you learned. They called my name and asked me to explain how I would repair some things using leather, repair of cushion seats and leather horse equipment. When I finish my examination, the director or commandant of the school, General Churchill told me that my presentation was the best that he had ever heard in his life. The General asked me if I would like to stay at Fort Riley. He said he would like for me to be stationed there. I told him that when I had called my company commander. I told him that I would be finishing school and he said for me to return to my unit. General Churchill then said, if your commander told you to return to the unit, then that's what you should do. However, if you ever want to come back here, just let me know. I was one of three black soldiers in a predominantly white class of soldiers. Even though the army was segregated, we stayed in the same quarters and attended class together and sat in the same class rooms. Since I was assigned to headquarters troop, they would give me the guidon to carry when we were marching, 87 whites and three blacks together. I would be leading the formation along with the first sergeant.

I returned to my unit and shortly after my arrival we left for a large maneuver in Nogales, Arizona. We rode up and down the creeks at night. The water was clear. In Nogales, one side of the street is the U.S. and the other side, Mexico. The street was called Pennsylvania Avenue.

When we went back to Fort Huachuca there was a pistol match scheduled. We used the 45 mm automatic pistol. I also could shoot very well the 37 mm rifle. I could always hit the target. When I was a child, I had a B-B gun and would go into the mountains to hunt. But in the army I had to shoot in accordance with army regulations. One day on the range, I asked my Captain if I could shoot in a position that I liked. I told him that I could tear that target up shooting my way. The captain told Sergeant Nelson to permit me to use my own position to shoot. I got down on my knees like this, and aimed at that target and hit the bull's eye. Everything went just fine and the captain was surprised.

On June 25, 1930, I received orders to report to Fort Myers, Virginia. When we arrived the all white Third U.S. Cavalry Regiment and a white artillery unit were stationed there. Within a week we were doing the same duties that the white cavalry unit were performing. We drilled on the large parade field at Fort Myer and used the riding hall. I had duty at Arlington National Cemetery at times. Many times, I would take enough ammunition to remain the entire day. The superintendent of the cemetery or my lieutenant would say, here comes another one, referring to the funeral processions. We would then prepare to fire the rifle volleys at the appropriate time. The cemetery was segregated, but we were still detailed to fire the volley for all funeral procession, black or white. The bugler and pall bearers were white. I could play the bugle. I was a bugler for seven years. I could play taps and all the calls. When we were on long marches,

I would blow the calls and also use hand signals for the commands. Mount up, and stop or halt. I normally would ride ahead of the main body, at least 5 miles sometimes. Our horses also knew what to do. They knew the signals better than some soldiers. If you were out somewhere in a mock battle, you could tap the horse on its knee and he would lay down immediately, then you could fire across his body. I would always carry an extra covering or blanket for my horse. Because when it rains a horse's ears should be protected. If rain gets in their ears, it causes an irritating sensation or feeling. When the weather was bad, we would ride in the riding halls. We also made motion pictures in the riding halls.

We performed for the late Emperor Haile Selassie of Ethiopia. We were scheduled to perform for him on another visit to the United States but he was ill. Therefore, we performed for his nephew Ras Desta, in the Fort Myer large Riding Hall. Our officers were sitting erect in the stands and other people were watching the Tenth Cavalry member perform using their horses. We made a picture like model of a ship on water. The men using their horses had a man standing on each others shoulders forming a very high level. Ras Desta was very impressed. Because later he left the stands and just looked, probably trying to understand how we were able to do the act. After we had dismounted, Ras Desta approached each man, asked their name and where they were from in the United States. There were three men from Howard University acting as translators for the Ras. However, Ras Desta could speak English very well and he then asked was anyone from Chicago, Illinois? Several men raised their hands. Ras Desta said very fine, because I went to school in Chicago. He also said that the performance was the best he had ever seen. The soldiers of the regiment were very proud to see and perform before an African royalty family member. The next visitor we performed for was Queen Mary of England. We had to meet her at the Washington, D.C. Union Station and escort her to the White House. Later, we gave a performance for her at Fort Myer. The men would remark about her beauty. She was small at that time and a beautiful queen.

My family and I came to Washington, D.C. in 1931. I was assigned to Fort Myer, Virginia from 1931 until 1939. Although I was detailed for a brief period at Fort Meade, Maryland. When I returned to Fort Myer, I was relieved by my brother-in-law, Sergeant William B. King, a Buffalo

Soldier of the Tenth U.S. Cavalry Regiment. When I returned to Fort Myer on January 1, 1939, I received orders to report to Fort Leavenworth, Kansas, to assist in the remolding and rebuilding of the Tenth Cavalry Regiment. I was assigned as a stable sergeant. I was in charge of the horses. We trained the new soldiers how to ride using the post riding halls in the winters.

In 1941, we played a game of baseball in Kansas City, Kansas in a baseball park near the stockyards. When we finished our ball game, we spent the night there. We had loaded our horses on a cattle car and we were heading for maneuvers in Lousiana. During a trip to Louisiana we stopped in a place in Missouri. The people said they had never seen soldiers. We spent a few days there, gave our horses a chance to exercise roam and graze, in open grass. We stopped again in Memphis, Tennessee. Then we stopped for one week in a town called McGehee, Arkansas. We camped out and checked our horses. There was a man in McGehee named Mr. Swann. They said he was one of the richest men in Arkansas, the entire state. He owned most of the real estate in town. Mr. Swann provided us with water, food and everything we needed. Later, his son joined the Tenth Cavalry Regiment. We had a truck assigned to each troop. We could ride a truck in the day or our horses.

We received orders to leave McGehee and proceed toward Shreveport, Lousiana. There was a contest during the travel. A group would leave from the east, called the Blues and the group from the west, called the Reds. The troop that arrived first was the winner of the Mock battle. Our troop would rest in the day and travel by night. We could make better time at night. The horses were trained to walk and trot, a nine mile per hour rate of travel. We would start out with a trot for 15 minutes, then rest 15 minutes, start again and continue trot, walk and rest. You will have made 9 miles per hour. My horse was a lead horse.

One time we went on a long march from Fort Myer to Pennsylvania, near Lancaster, Pa. The march formation consisted of the Tenth Cavalry, National Guard units from the District of Columbia and Maryland. There were white and black units marching in the formation. Rosslyn, Virginia was our assembly point. I had to position myself 3-5 miles ahead of the main body. Our first stop was Rockville, Maryland. We stayed there for 15 minutes to rest our horses. I can remember passing a large snake farm near Thurmont, Maryland. After we passed the snake farm, we camped for the night near a stream. We had our cooks and kitchen truck. The signal was given to bivouac and the men began to pitch their tents. The next morning as I was preparing to get my breakfast, I reached in my saddle bag to get my mess kit and a large snake came out. I threw the saddle bag with the mess kit all away. I have a friend who lives in Washington and everytime he sees me, he mentions that morning and the incident. He loves to kid me about it. You know the Cavalry soldiers always carry their riding gloves with them, especially in the field. I had my gloves on when I reached into my saddle bag. The next day we continued our march. We marched to Indiantown Gap, Pennsylvania and stayed there for a week. Then we proceeded east and stopped briefly at a place where the Amish people were selling things. We stayed there for several hours. The men were buying different goods that the Amish people were selling. We continued our march and stopped for the night at Gettysburg, Pennsylvania. That night we camped on the battle field near some large tomb stones. We fed our horses and they had some nice alfalfa and hay. Just as everyone had gone to bed for the night and it was quiet, suddenly you could hear a yelling or screaming in the air or as it appeared, the horses became uneasy and began to jump. Everyone witnessed the movement of the horses and heard the screams, yes, the men and the officers. The next day a man who was a caretaker at the battlefield for almost 50 years said, that everytime a group of soldiers camp out in that particular area, the screaming starts. I witnessed it and heard the sounds and so did other people that night.

After the maneuver in Pennsylvania, I was later assigned to Camp Funston, near Fort Riley, kansas. When the Japanese bombed Pearl Harbor in December 1941, we received orders to take the shoes off the horses, give them plenty hay, but no grain and place them in the corral. we were told to send all of our clothes and personal things home, or give them away. I sent a foot locker home.

General Benjamin O. Davis Sr. had a son who had graduated from West Point and he was assigned to the Cavalry. When I reported for duty, I was told to pack all my belongings and that I would be in charge of 82 men. I was introduced to Lieutenant (Lt.) Benjamin O. Davis Jr., and told to follow him and take orders from him. I was told that he would tell me what to do. I reported to Lt. Davis with 82 men.

Before I departed Fort Huachuca, I went to the corral to see the horses. When I called two of them, they came running to me and I patted them. There were some people who were watching. They said look how that man has control over those horses and how the horses follow him. I can remember one time when the horses were stampeding, no one could stop them. I just waved my hand and the horses stopped. Some officers were watching and could not believe it.

The 82 men and I went to Montgomery, Alabama and later to Tuskegee. At Tuskegee, we went into a large building where each man was given a motor that had been placed on a large block. Each man was to tear the motor down and rebuild it. The instructor told me that I would not be able to go up in the air, because I was too old. They were only taking men 17-36 years old for pilot training.

Later, I received an assignment to a place near Enterprise, and Dothan, Alabama. The Twenty-fourth Infantry Regiment was there and some other units. They were training recruits. There was a large pond where the recruits would be tested for their swimming abilities. I had my roster and I would sit down and have the soldiers go out to the pond in a little boat and observe how they could swim. Those who could not swim would be sent to a special class to learn how to swim. At this time, I was assigned to the Twenty-fourth Infantry Regiment. Later, I was sent to McCoy Field, North Carolina. There we were tested by going into a large drum. It would go high up in the air and would rotate around. The instructor told me that I had better nerves than the average young soldiers. Later, I was sent to the South Pacific.

I served in a special cavalry unit. We were organized in North America and first went to Panama to an area near the Bay Washington River. I was no longer a member of the Tenth U.S. Cavalry Regiment. We arrived on Saipan. Saipan was very beautiful at night. There was plenty of fruit, fresh strawberries, bananas, and pineapple. The Japanese had a strong hold on the large hills on the island. Tokyo Rose had told the Japanese people that the Americans would kill them and assault their women. Some of the people believed Tokyo Rose. They were prepared to commit "Hari Kari" (suicide). There were many Japanese snipers still on the island. We had two white lieutenants who were shot just after walking out of the mess hall. Many white soldiers were being killed by possible snipers.

When I returned to the United States from the South Pacific, I was told to take some men to Pittsburgh, Pennsylvania. I told the major he would have to get someone else, I said, I am retiring, I have 30 years. The major said can you prove it? I said give me 5 minutes I went and got my papers. The major was surprised and he called in some other military people and said, look at this man, he looks like he is 20 years old and he has served 30 years. He then told me that he could not change my orders and that I had to go to San Francisco, California and that there was a man there who was in charge of all soldiers from San Antonio to the Philippines. I got my bag at 10 o'clock in the morning and went to San Francisco. I received my orders and retired after 30 years of service around 1947. I retired the second time from a civilian job as a security officer at the National Institutes of Health.

RESPONSES TO SOME DIRECT QUESTIONS

DURING THE INTERVIEW

Some activities that I participated in were the band and sports. When I was in the band, I reported to Warrant Officer Hammond. We spent at least an hour for practice. Hammond would come into the room and tap on the table, and raise his hand. We knew he was ready to begin. He would tell us what key we needed. Whatever you were playing, you had to be ready. I also enjoyed playing baseball, long distance runner in track and polo. Everything that would be available, I participated. I was also a lightweight boxer.

Now on Poncho Villa, I never understood him. He was raiding everywhere. Some people would say that Poncho Villa was not doing all those raiding. People would cross the line, take a large amount of cattle and cross back into

Mexico and say it was Poncho Villa. They said Villa had a large farm. I learned later that it was one of Poncho Villa's men that shot him.

I met some Indian scouts when I was stationed at Fort Huachuca, Arizona. They had moved 35 families from Fort Apache reservation, around 1928-1929. We sent trucks for them, however, we still had some wagon trains. The Indian scouts were given quarters on post. But the scouts and their families did not rest so good in the buildings. They went to the colonel and asked to leave the buildings. They said their children sleep better in the tepees. An area was cleared for them, and a large tent area provided for them. They erected their tepees, and had rugs on the ground. Their children were very happy.

There were some 50 Indian scouts with the Tenth Cavalry. Four scouts were assigned to each troop. We had an Indian scout named Lonnie, he was in charge of our garden. The garden was located on post and we raised vegetables and also had some cattle. An Indian scout assigned to my troop was Chow Man. We also had four Indian scouts in our band. I can remember when the scouts would bring their families to our big dinners.

The Indian Apache scouts would have a "Maiden Head" ceremony. When a girl would reach a certain age and be eligible for marriage. The older women would prepare a padded bed decorated with flowers. They would get a large pitcher and make a tea like drink, called "tunapot". The Indian males would paint themselves red and start beating the drums. The senior Indian squaws would stand beside the girl and start to patting her and then they would hop around and sing. Everyone would join in, hopping and singing. One time, I joined in and we all had a fine time.

There were times we would be marching along with all white units. Sometimes we would be marching on maneuvers with the Seventh and Eighth Cavalry regiments. They came from El Paso, Texas and we came from Kansas. Some days during a march, there would be a white platoon and black platoon together. One time we marched from Fort Huachuca to El Paso, Texas, 279 miles. The white cavalry and black cavalry together, if nobody looked at them, they were the same as one, so proud to be cavalrymen.

We marched in President Franklin D. Roosevelt's parade with whites in 1939, February 2. That was Roosevelt's second inaugural. When I went to Fort Leavenworth, Kansas and received my horse, I named it "Franklin D". He was a tall horse.

Now I met my wife when I was stationed at Fort Huachuca, Arizona. I went to a rifle-pistol competition in Douglas, Arizona. I did not want to stay in a hotel because I needed plenty space to clean and prepare my weapons for the competition. I knew how to clean an entire weapon. I had worked in the weapon store room. I went to a barber shop and asked about a place to stay. The barber told me he knew some nice people on Kent Avenue. he called them and when I arrived at the home, there was another soldier there that I knew. There was a man and two women present in the home. I looked around and saw someone standing in the door. When they saw me looking they moved back. I asked who was that person. Her brother said, that is my sister. I said the last time I saw her she was very small. He said she is in high school now. He called his sister in for me to meet her. He said, this is Private Matthews and he will be participating in the competition. He is a member of the rifle and pistol team. I invited the young lady to the competition. But she said, that she would be in school all week, but would be off Saturday. I said, if you can come out on Saturday you will get a chance to see me participate. On Saturday, she, her mother and brother came to the competition. I had a nice place for them to sit and observe the rifle and pistol competition. After the meet, I went back to Fort Huachuca.

When school vacation was over, this nice young lady that I met whose name was Genevieve made a visit to Fort Huachuca. She was on school vacation and made the visit with her girl friend whose father was a post barber. Genevieve's mother called and wanted to know who was her daughter's escort. I told them I was. I told her mother I would be her daughter's escort. They wanted to know in case anything would happen, they wanted to know who to call. I showed Genevieve around the post, and took her to parties. We talked about many things and I guess people would call it courting. We corresponded for about 3 years. She wanted to finish high school. When Genevieve finished high school we got married on April 19, 1929.

I was married to my wife for 52 years until her death. I remember my wife as a person who was quiet and did not have a lot of talk with everybody. She was a good listener. We always discuss things. If anything went wrong, we would sit down and discuss the problem. We did this for 52 years. We never had much of an argument. We talked things over and we would learn to agree with each other. She was a friendly person, a good cook and a member of the Eastern Star, America's Legion (Woman's Auxiliary). Everybody liked her. She was also active in club work and church activities. My wife's grandmother was an Indian and her grandfather was a free black man. They had always lived in Arizona, in Douglas and Nogales. Her step father was a member of the Tenth Cavalry Regiment. He was Marcus Nelson. When my son was born in 1934, my wife integrated Walter Reed's Army hospital. When she arrived at the hospital, all of the white women were on one ward. The black women's maternity section was on a porch like ward. Her white physician asked her why was she in the special area? He was told that she wanted to be in a private area. The white personnel in charge tried to hush it up. The doctor said he was not going to have this arrangement. My wife had told the doctor that it was not the way, the people had told him, it was because she is a Negro women. Then the doctor had her moved inside where the white women were located.

When I look back at segregation in those days, I realize that it was pretty hard at times. We went to different places and sometimes were treated pretty nice. We would go out and do certain things that other blacks could not do. When I was stationed at Fort Myer, Va., I participated in horse shows with my horse "Kansas Red". They were all white horse shows at Barry Farms and a place on East West highway. I would stand Kansas Red where people could see him standing so straight. The people would ask questions and wanted to know how he could stand there just like a human being, so erect. My brother-in-law and I took part in horse jumping and high hurdle competition in white horse shows. I have my cups to show some of the events I won. Yes, we did some things that black civilians were not able to do during those segregated days.

COMMENTARY

The Saga of Mark Matthews is an outstanding contribution to American civilian and military history. First Sergeant Mark Matthews celebrated his 100th birthday on Sunday, August 7, 1994. His 100 years have been most rewarding and productive. He also found the time for membership in organizations such as the Elks, Masons (33rd Degree), National Pythians, and American Legion. He was an athlete, demonstrating his superb abilities as a pole vaulter, long distance runner, lightweight boxer, rifle and pistol team member and polo team member.

The pictures that follow are representative of his illustrious career as a Buffalo Soldier and his proud and loving family and extended family members. I salute you again First Sergeant Mark Matthews and may God continue to bless you with more healthy and fruitful years in your well deserved retirement years of long life.

Robert E. Greene
August 7, 1994

A PICTORIAL VISIT

MARK MATTHEWS

AND

HIS EXTENDED FAMILY

First Sergeant Mark Matthews

Genevieve Hill Matthews

Genevieve Matthews' grandparents
Mary and Henry Tibbs

**A Real Buffalo Soldier
First Sergeant Mark Matthews**

**First Sergeant Mark Matthews
(Far right)**

First Sergeamt Mark Matthews on his horse Kansas Red

The Real Buffalo Soldiers 217

First Sergeant Mark Matthews
Far Left, Fourth Row
1926 training course - Fort Riley, Kansas

First Sergeant Mark Matthews

The Real Buffalo Soldiers 219

First Sergeant Mark Matthews

**President Clinton
greets
Two Real Buffalo Soldiers**

Mark Matthews' brother-in-law
Sgt. William B. King - Tenth Cavalry Regiment

Sergeant William King

**Machine Gun Troop, Tenth Cavalry
Fort Myer, Virginia
Captains, Lieutenants, Sergeants and Corporals**

**Machine Gun Troop, Tenth Cavalry
Fort Myer, Virginia**

Sergeant William B. King
Second Row Left
Mark Matthews' brother-in-law

Mark Matthews
A member of Security Force
National Institutes of Health
Bethesda, Maryland

Mark Matthews and his friend
at his retirement party
National Institutes of Health

Mark Matthews' retirement party

Mary Matthews Watson
Mark Matthews' daughter

Shirley Ann Matthews Mills
Mark Matthews' daughter receives an award

Lance Corporal Michael Anthony Watson
Mark Matthews' grandson

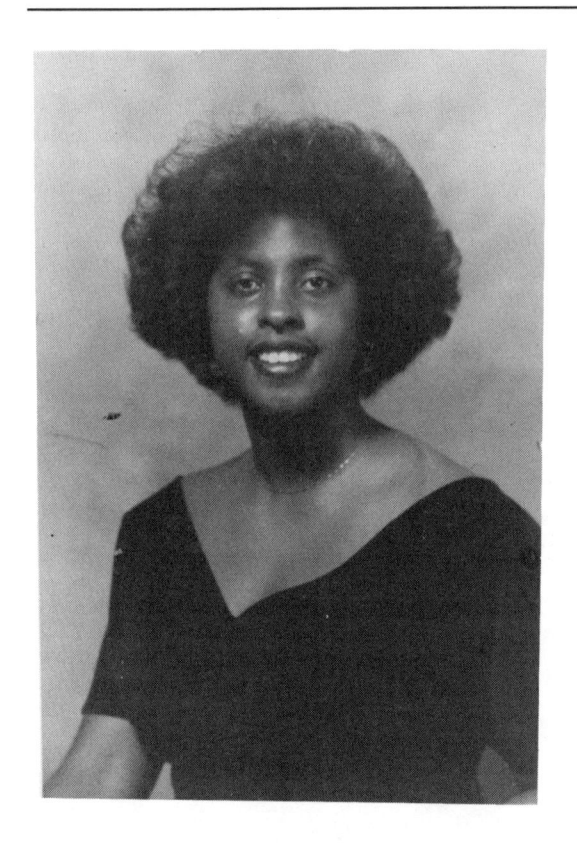

Mark Matthews
granddaughters

Andrea L. Edwards

Debbie C. Jackson

Michael Anthony Watson Jr.
great grandson

Mark Matthews'
great grandchildren

Sheila and Wallace

Mary Watson and
great grandchildren
Left to right
Andrea, Bryan
and Shari

Mark Matthews'
great grandchildren

Dasmine

Shelby, Danielle and Shannon
Left to Right

CHAPTER 13

BUFFALO SOLDIERS A COMMENTARY

BUFFALO SOLDIERS

By Major Clark

According to a legend of the Old West, Black regular army soldiers involved in the Indian Wars were first called "Buffalo Soldiers" more than a century and quarter ago. However, almost from the beginning, historians have disagreed concerning certain version of the legend, especially with regard to which units were so named; and the meaning of the name.

I was first introduced to the Buffalo Soldier controversy more than 50 years ago, in August 1940, when I enlisted in the Army and was assigned to the newly activated Three hundred forty-ninth Field Artillery regiment at Fort Sill, Oklahoma.

Most of the cadremen for the Three hundred forty-ninth had been furnished by the four black regular army units that had been involved in the Indian Wars and the Spanish-American War; two cavalry regiments (the Ninth and Tenth) and two infantry regiments (the Twenty-fourth and Twenty-fifth). These cadremen had enlisted in the Army some time between the Spanish American War and World War I. We called them "old soldiers" because they were in their middle or late forties and nearing eligibility for retirement.

Early in 1941, the Three hundred forty-ninth furnished a cadre to activate the Forty-six Field Artillery Brigade at Camp Livingston, Louisiana. Included in the cadre were "old soldiers" like Master Sergeant William Harrington, who passed away in 1994 at the age of 99; and relatively new soldiers like me. About two months later, after I became a technical sergeant (one stripe less than a master sergeant), I began to associate with some of the old soldiers, and was present for a number of spirited discussions about the past history of their former cavalry and infantry units.

Although most of their accounts included the legend that, during the Indian Wars, the Indians called their predecessors Buffalo Soldiers, few of their accounts were specific concerning which Indian nations or army units were involved, or where or when the first such event occurred.

One of the old soldiers was a student of military history, and was less emotional than the others about the past history of their units. He deplored the fact that there was so much controversy about a name based on a variable legend.

One area of controversy related to the question concerning which units were called Buffalo Soldiers by the Indians. For example, some former member of the infantry units contended that all of the colored regular army soldiers involved in the Indian Wars were called Buffalo Soldiers by the Indians. On the other hand, some former members of the Ninth and Tenth Cavalry believed that the only soldiers qualified to be called Buffalo Soldiers were those who had served in the cavalry.

Some former members of the Tenth Cavalry were even more exclusive. They wanted to limit the name Buffalo Soldiers to the Tenth Cavalry because it was the Tenth Cavalry to which the term Buffalo Soldiers was first applied.

Although all four regiments had been involved in the Indian Wars, there was no reference to Buffalo Soldiers in the official documents pertaining to the coat of arms, regimental badge or distinctive insignia of either of the other colored regular army regiments. Only the Tenth Cavalry featured a buffalo on its regimental device and distinctive badge approved by the army.

A second area of controversy related to the reason why they were called Buffalo Soldiers by the Indians.

In that regard the legend had at least three variations. They were called Buffalo Soldiers by the Indians: (1) because they fought with the fierceness of a corned buffalo; (2) because they resembled buffaloes when they were dressed from head to toe in huge robes made from buffalo hides to protect themselves from the harsh, cold, blustery winters on the plains; (3) because the wooly heads of the colored soldiers were so much like the matted cushion between the horns of the buffalo.

A third area of controversy related to the assumption that, because the buffalo was considered a sacred animal by the Indians, "Buffalo Soldiers" was a term of respect. Many of the soldiers had been involved in combat as a part of the 92nd Division during World War I and, although they had heard that some of their generals spoke of some enemy generals with terms of respect, it was contrary to their experience that soldiers at the combatant level spoke of enemy soldiers trying to kill them with terms of respect.

A fourth area of controversy related to the questions of whether all colored soldiers who served in one of the regular army units at some time from 1866 until then (1941) should be called Buffalo Soldiers without further distinction.

It was the old soldiers's opinion, that before getting involved further in the fourth controversy, a truce should be declared in regard to the first controversy, and the version of the legend should be accepted which states that all of the colored regular army units were called Buffalo Soldiers by the Indians at some time. Then, to avoid confusion, the Buffalo Soldiers could be classified according to the period during which they served as either "Original Buffalo Soldiers", "Transitional Buffalo Soldiers," "Associate Buffalo Soldiers, or "Traditional Buffalo Soldiers."

"Original Buffalo Soldiers" were those who had served in one of the units during the period of 1866 until 1900 which included the Indian Wars and the War with Spain.

During the Indian Wars, favorable treatment and perception of proficiency of the colored soldiers by the white citizens was directly proportional to the number of hostile Indians in the area.

During that period, those units gained a reputation as proficient troops and earned many first class honors. This reputation continued until after the Spanish American War. Accordingly, for Black Americans at the beginning of the twentieth century, the name Buffalo Soldiers symbolized brave, proficient soldiers.

"Transitional Buffalo Soldiers" were those who had served in one of the units during the period of transition from 1900, when adverse changes began in the first class honors received, and the perceived proficiency of the units, especially the infantry; until the end of World War I in 1918, when the adverse changes had also affected the cavalry units.

An soon as some of the units were transferred from frontier stations to stations in the South, those units were subjected to unfavorable treatment and unfavorable perception of their proficiency as soldiers. They did not receive any more first class honors.

1906. Brownsville, Texas, incident involving the Twenty-fifth Infantry resulted in the discharge, without honor, of Companies B, C, and D of that regiment.

1916-1917. Mexican Punitive Expedition involving the Tenth Cavalry, which was considered to be a failed operation.

August 1917. Riots in Houston, Texas, involving the Twenty-fourth Infantry. Early in December 1917, thirteen men involved in the Houston riots were sentenced to hang. They were hanged four days later.

The regular army Buffalo Soldier units were not involved in World War I combat overseas.

An incident took place at the end of the transitional period for which the old soldier had no satisfactory explanation. In ad Editorial in the *CRISIS* for September 1918, Dr. William E. B. DuBois listed the wars, prior to World War I, in which Negro soldiers had participated, and listed the benefits received by Negroes following that participation. The editorial stated:

> Five thousand Negroes fought in the Revolution...At least three thousand Negro soldiers and sailors fought in the War of 1812...Two hundred thousand Negroes enlisted in the Civil War... Some ten thousand Negroes fought in the Spanish-American War...

Why did Dr. DuBois fail to list the participation of the Buffalo Soldiers in the Indian Wars during which they gained their reputation as proficient soldiers and earned more first class honors from the U.S. Army before or since.. In October 1917, the World War I 92nd "Buffalo" Division was organized.

At first, because of the symbolism associated with the name, the regimental commander of the Three Hundred Sixty Seventh Infantry called his regiment "The Buffalos." Since the regular Buffalo Soldier units were not involved in World War I combat overseas, these soldiers provided for the continuity of participation of a Buffalo Soldiers unit in all of our Nation's wars since the Civil War. Later the name was adopted by the entire Ninety-second Division.

Afterwards, the army approved a buffalo shoulder patch for wear on the uniforms of the division's 25,000 men. This occurred four years before the army approved the Tenth Cavalry's regimental distinctive insignia that included the likeness of a buffalo. "Traditional Buffalo Soldiers" were those who began their service after World War I in one of the regular army units that perpetuated the Buffalo Soldiers legends and traditions.

In October 1942, I became an "Associate Buffalo Soldier" when I was commissioned as an officer and assigned to the World War II Ninety-second "Buffalo" Infantry Division. This was the first unit in history recognized by the army as a Buffalo Soldier unit at the time of its activation.

I found out immediately - and was reminded every day for the next three years - that it was a historical fact, not a legend, that we were called Buffalo Soldiers by the army. Every member of the division was required to wear a "buffalo" shoulder patch; the division newspaper was called "The Buffalo"; and the division kept a live buffalo as a mascot until we left the U.S. for combat service overseas.

The Ninth and Tenth Cavalry Regiments were inactivated in 1944 and were not involved in World War II combat. The Twenty-fourth and Twenty-fifth Infantry Regiments were involved in combat during World War II, both in the Pacific Theater; the Twenty-fourth as a separate regiment; the Twenty-fifth as an organic regiment of the Ninety-third Division.

The Twenty-fourth Infantry was the only Buffalo Soldier unit involved in the Korean War. It was inactivated in 1951 while the Korean War was still in progress.

Some recent events have gained greater public attention because of the promotional activity incident to the dedication of the Buffalo Soldier Monument at Fort Leavenworth, Kansas; the designation of July 28, 1992, as Buffalo Soldiers Day; and the issuance of the Buffalo Soldiers Postage Stamp. Unfortunately, these events have tended to further confuse rather than resolve the controversy.

In one area of confusion, the Buffalo Soldier classification system proposed by the old soldier in 1941, would be helpful.

Within the past year, two prominent buffalo soldiers have died. In the 20 September 1993 issue of a national magazine, it was reported that "Jones Morgan, *the last U.S. Buffalo Soldier,* Dies." Seven months later, in the 11 April 1994 issue of the same Magazine, it was reported that "Retired Sergeant Major William Harrington III, *one* of the last remaining Buffalo Soldier, died only a few months before he would have reached the age of 100."

Many individuals across the country - some forty years younger than Morgan and thirty years younger than Harrington - have attempted to exploit the publicity concerning the Buffalo Soldiers by claiming to be one of the *few remaining Buffalo Soldiers*.

In another area of confusion, there appears to be little hope for clarification. For example, one of the few specific accounts concerning the origin of Buffalo Soldiers was used by Senator Nancy Kassebaum of Kansas in her speech on 12 March 1991 introducing SJ Res 92 to designate July 28, 1992, as "Buffalo Soldiers Day." According to that account, the event followed an attack by the Cheyenne warriors on "F" Company of the Tenth Cavalry while it was on patrol near Fort Hays, Kansas on 2 August 1867. After the battle an army scout overheard the Indian speaking with respect about this first encounter with black soldiers. The warrior called them Buffalo Soldier because they fought with the fierceness of a cornered buffalo.

In his speech in July 1992 dedicating the Buffalo Soldier Monument at Fort Leavenworth, Kansas, General Colin Powell stated:

The Buffalo Soldiers were not the only ones in the struggle. The Twenty-fourth and Twenty-fifth Infantry Regiments, the 92nd and 93rd Infantry Division... and thousands of other brave Black Americans have gone in harm's way for their country since the day of the Buffalo Soldiers...

By placing the Twenty-fourth and Twenty-fifth Infantry and the Ninety-second Division in the category with units other than Buffalo Soldiers, General Powell seems to indicate that he considers only the two cavalry regiments (the Ninth and Tenth) to have been "Buffalo Soldiers." On the other hand, the promotional material distributed by the Buffalo Soldiers Educational and Historical Committee during that period stated that the Buffalo Soldiers were members of the Ninth and Tenth Cavalry and Twenty-fourth and Twenty-fifth Infantry Regiments.

The "Associate Buffalo Soldiers" were excluded from consideration as Buffalo Soldiers by Senator Kassebaum, General Powell, and the Buffalo Soldiers Educational and Historical Committee. As a matter of fact, when the promotional activities concerning the other Buffalo

Soldiers were beginning to accelerate, a Department of Defense action had the effect of excluding the World War II Ninety-second, "Buffalo" Infantry Division, the only unit recognized as a Buffalo Soldier unit at the time of activation, from being considered in any capacity. The revised edition of the Department of Defense Booklet, *Black Americans in Defense of Our Nation* (which summarizes African American military contributions since the Revolutionary War) omitted the World War II Ninety-second Buffalo Infantry Division.

After that edition was published, I wrote to General Powell, then Chairman of the Joint Chiefs of Staff, informing him of the omission.

The response was:

> *Thank you for your loyalty and dedication to what was an outstanding group of soldiers in World War II, the 92nd Infantry Division.*
>
> *The Story of the 92nd Division was not intentionally omitted form the current edition of Black Americans in Defense of our Nation. The omission was an inadvertent error that the authors assure me will be corrected in the next edition.*

It is three years later, General Powell has retired, and we are still waiting for the correction.

In the meantime, the television documentary, "Buffalo Soldiers, The Legend Continues" was presented on the Arts and Entertainment Television Channel on 28 February 1994, and later on the Educational Channels across the nation... Although a ten minute segment of the documentary was devoted to the fifty men of Seminole Negro Indian Scouts, only a five minute segment was devoted to the almost 14,000 men of the World War II 92nd "Buffalo" Infantry Division which, during eight months of combat operations against the "master race" overseas, sustained over 3000 casualties (killed, wounded or missing) and earned several thousand decorations.

Fifty-two "Associate Buffalo Soldiers" belonging to the 365th Infantry (Ninety-second Division) were killed while carrying out a successful special operation during the second week of February

1945. The number killed within that one week in the 365th was more than the total number of men assigned at any one time to the Seminole Negro Indian Scouts. However, the 365th Infantry was not mentioned in the documentary.

SPIRITUAL DESCENDANTS OF BUFFALO SOLDIERS

However, all is not lost. While being denied their own category, "Associate Buffalo Soldiers" can fit into another category.

At the dedication of the Buffalo Soldier Monument, General Powell introduced a category of Buffalo Soldiers which seems to include all African American Soldiers who have served honorably in combat:

We are not here today to criticize America of 150 years ago but to rejoice that we are in a country that has permitted a spiritual descendant of the Buffalo Soldiers to stand before you today as the first African-American Chairman of the Joint Chiefs of Staff.

Major Clark
Lieutenant Colonel, AUS Retired

Lieutenant Colonel Major Clark enlisted in the U.S. army on 31 August 1940. He was assigned to the first black field artillery unit activated in the regular army in the history of the United States, the 349th field artillery at Fort Sill, Oklahoma. As an enlisted man, Clark advanced through the grades from private in 1940 to technical sergeant in 1942.

In 1942, he became a commissioned officer in the field artillery after graduating from the Field Artillery Officer Candidate School at Fort Sill, Oklahoma. As an officer, Clark advanced through the grades from second lieutenant in 1942 to lieutenant colonel in 1957. He retired in 1960 after twenty years of active duty.

Some of Major Clark's significant assignments have been field artillery battery commander, battalion intelligence officer, and historian of the 597th field artillery battalion during World War II, assistant professor of military science and tactics, Hampton Institute, Virginia. Senior artillery advisor to the 3rd Korean army division during the Korean War, battalion executive officer of 595th field artillery battalion at Fort Sill, Oklahoma and sixty ninth anti-aircraft artillery battalion, New York, and army general staff officer in the Pentagon.

Clark is also a graduate of the 1953-1954 artillery officer advanced course at Fort Sill, Oklahoma and Fort Bliss, Texas, the command and general staff officer course (1955), the special (nuclear) weapon course (1956), and the senior officer nuclear weapons employment course (1958), all at Fort Leavenworth, Kansas. Colonel Clark is the recipient of the Bronze star, Italy, 1945, the Oak Leaf Cluster to the Bronze star and the Changmu medal (Korean) with Gold Star in Korea, 1952, the Army Commendation Medal, Pentagon, 1960 and numerous service medals and special awards. In 1980 Clark was inducted into the Field Artillery Officers Candidate School Hall of Fame at Fort Sill, Oklahoma

CHAPTER 14

PICTORIAL REFERENCE OF THE REAL BUFFALO SOLDIERS

FOREWARD

Because of my interest in military history and especially the exploits of the "Real Buffalo Soldiers, I cannot forget those racist remarks and attitudes of a white artist who could never see a positive scene or conduct an interview with some black soldiers who were fortunately able to receive an education so they could articulate normal English. However, Remington's attitude and intentions of that day are still present by some honorable and possibly innocent minded Caucasians and other non-blacks who feel that the man on the street or the so called common black's characteristics are representative of most blacks.

I deeply believe that the selected pictures in this manuscript will provide the reader with a true factual representation of the physical and descriptive biological genetic diversity of African Americans.

Cadet
Henry O. Flipper

Chaplain
T.G. Steward

Sergeant John Denny

Corporal Isaiah Mays

Sergeant Henry Johnson

Sergeant Thomas Shaw

Colonel John W. Huguley III
Military Surgeon
Great Grandson of Buffalo Soldier
Sgt Thomas Shaw

Sergeant Brent Woods

**Cadet
John Alexander**

**Lieutenant
John Alexander**

Henry Ossian Flipper

**Lieutenant Powhattan H. Clark
Rescues Corporal Scott
Tenth Cavalry, 1880**

Musician Miles Monroe

Sergeant William H. Tompkins
Tenth U.S. Cavalry Regiment
Medal of Honor Winner
Spanish American War, 1898

The Real Buffalo Soldiers   265

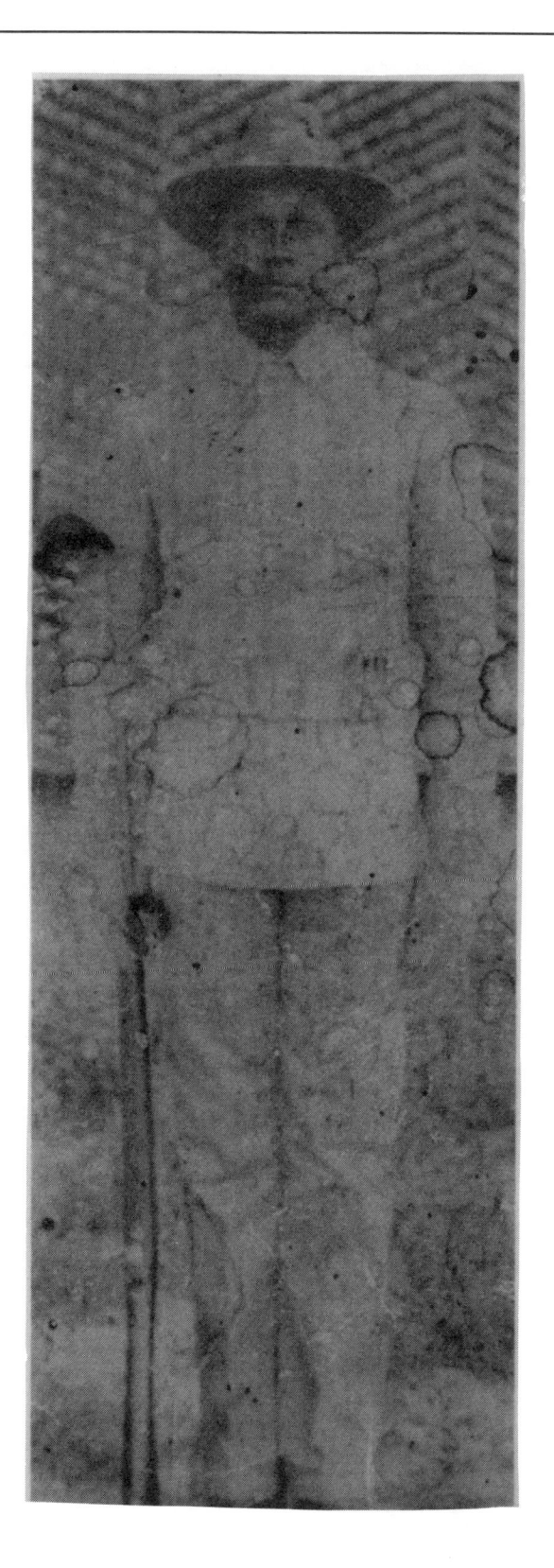

**Seminole
Negro Inidan scout
Billy July**

266   **The Real Buffalo Soldiers**

**Seminole Negro
Indian Scout
Bill Williams**

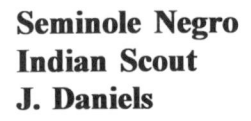

**Seminole Negro
Indian Scout
J. Daniels**

**Seminole Negro Indian scout
Charles Daniels and Family**

Seminole Negro Indian Scouts

John Jefferson

**Hampton Institute
Presence of Indian Students**

**Cecelia Spears Green
African, Seminole Indian and
European Descent**

**Wilma Belle Green Macer
African, Seminole Indian
and European Descent**

Chaplain
William H. Anderson

Sergeant Major
Edward L. Baker Jr.

274   The Real Buffalo Soldiers

LTC Allen Allensworth

George Washington Williams

Corporal George G. Anderson

Chaplain William T. Anderson

Lieutenant John Buck

Sergeant Horace W. Bivins

Lieutenant Arthur M. Brown
Surgeon Tenth U.S. Cavalry

**Surgeon Arthur M. Brown
Tenth U.S. Cavalry Regiment**

Chaplain Louis Augustus Carter

Brigadier General Benjamin O. Davis Sr.

**Second Lieutenant
Saint Foster**

**Chaplain
Oscar J.W. Scott**

Bandmaster Wade H. Hammond

**Regimental Color Sergeant
Abraham Hill
Twenty-fourth Infantry Regiment**

Lieutenant J. H. Hill

Private Johnson
Company E, Twenty-fifth Infantry
1869-1870

Bandmaster Walter Loving

**Sergeant
William McBryar**

Marion C. Rhoten

First Lieutenant Edward A. McDowell
Surgeon, Tenth U.S. Cavalry

Thomas Samuel P. Miller

Lieutenant Joseph M. Moore

Second Lieutenant John C. Pendergrass

Chaplain Henry V. Plummer

**Chaplain
George W. Prioleau**

**Lieutenant
Joseph H. Johnson**

Lieutenant John C. Proctor

First Sergeant John C. Sanders

Lieutenant A. J. Smith

**First Lieutenant
Jacob C. Smith**

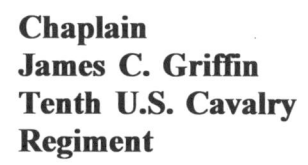

**Chaplain
James C. Griffin
Tenth U.S. Cavalry
Regiment**

Sergeant Major Charles B. Turner

Lieutenant William Washington

Color Sergeant
William C. Wilcox

Regimental Sergeant Major
Walter B. Williams

Colonel Charles Young

Colonel Charles Young

**General Pershing, General Bliss
and Lieutenant Colonel Young**

**Tenth Cavalry Regiment
Former Prisioners at Carrizal,
Mexico, 1916**

General John "Blackjack" Pershing

**Commander Wesley Brown, First Black Graduate
U.S. Naval Academy - Stands Beside Georgia Roadmarker
Honoring Henry O. Flipper, First Black Graduate
West Point Military Academy**

Chaplains of Color

**A Spanish Blockhouse on
San Juan Hill, Santiago, Cuba, 1898**

Color Guard, Ninth Cavalry Regiment

Ninth Cavalry Regiment

**Twenty-fifth Infantry Regiment
Fort Snelling, Minnesota, 1888**

**Twenty-fifth Infantry Regiment
Fort Shaw, Montana, 1888**

**Twenty-fifth Infantry Regiment
Fort Missoula, Montana, 1895**

**Twenty-fourth Infantry
Church Service, Manila, Philippine**

**Twenty-fifth Infantry Squad Room
Fort Huachuca, Arizona, 1895**

**Twenty-fifth Infantry Regiment
Baseball Team, 1903**

**NonCommissioned Officers Staff and Band
Fort Lawton, Washington, 1912**

Court Martial of Twenty-fourth Infantry Soldiers

Tenth Cavalry Regiment Hospital Corps

**Far Right. Dorsie Willis
Veteran, Twenty-fifth Infantry
Brownsville Affair, 1906**

**Tenth Cavalry Regiment
Color Sergeants**

The Buffalo

**Tenth Cavalry Regiment
Trophies and Awards**

The Real Buffalo Soldiers 325

Trumpeters, Tenth Cavalry Regiment

**Surgeon
William C. Warmsley**

**Tenth U.S. Cavalry Regiment
Staff and Band**

**Tenth U.S. Cavalry 1914
Noncommissioned Officers**

**Tenth Cavalry Regiment
Baseball Team 1902**

**Baseball Team
Troop M, Tenth U.S. Cavalry
1908, Fort Riley, Kansas**

Present Saber
Machine Gun Troop
10th Cavalry
Fort Myer, Virginia, 1931

**Tenth Cavalry Regiment
"The Hunt" 1934
Fort Leavenworth, Kansas**

**Tenth Cavalry Regiment
Polo Team**

The Real Buffalo Soldiers 333

Tenth Cavalry Machine Gun Troop

Tenth Cavalry Machine Gun Troop

Tenth Cavalry Machine Gun Troop

Tenth Cavalry Machine Gun Troop

Tenth Cavalry Regimental Color Guard

**Tenth Cavalry Regiment
Pass In Review**

**Tenth Cavalry Regiment
Pass In Review**

Tenth U.S. Cavalry Band

**Tenth Cavalry Regiment
Mounted Drill**

**Tenth Cavalry Regiment
On A March Advance Party**

The Real Buffalo Soldiers 343

**Tenth Cavalry Regiment
On A March**

**Tenth Cavalry Regiment
Fort Custer, Montana**

**Tenth Cavalry Regiment
In The Field**

**Tenth Cavalry Regiment
In The Field**

Tenth Cavalry Regiment Trooper

**Tenth Cavalry Regiment
Remount Training**

**Tenth Cavalry Regiment
Remount Training**

**Tenth Cavalry Regiment
Troopers**

**Tenth Cavalry Regiment
Trooper**

The Real Buffalo Soldiers 351

Tenth Cavalry Regiment
Regiment Training

**Tenth Cavalry Regiment
Light Machine Gun Training**

**Tenth Cavalry Troopers
With Light Machine Gun**

Tenth Cavalry Regiment Soldier

**Tenth Cavalry Regiment
Horse - Shoeing shop**

**Tenth Cavalry Regiment
Auto Repair Shop**

**Tenth Cavalry Regiment
Auto Shop**

358   The Real Buffalo Soldiers

**Tenth Cavalry Regiment
Field Mess**

**Tenth Cavalry Regiment
Dance Band**

Tenth Cavalry's Mess Hall

Buffalo Soldiers Escort Stage Coach

Tenth Cavalry Mess Hall

**Buffalo Soldiers
Tenth U.S. Cavalry Regiment**

Tenth Cavalry Troopers

APPENDICES

APPENDIX I

They Did not Tell Me

Unfortunately, in our multiracial society some minorities' rich history has been and even today is missing in many popular scholarly manuscripts of American history. The peoples of color or African American's history is still missing from the pages of the elementary, secondary and college and university's major American history textbooks. The following true facts concerning the Real Buffalo Soldiers of the Ninth and Tenth Cavalry Regiments and Twenty-fourth and Twenty-fifth Infantry Regiments are facts that they did not tell me in the textbooks and probably are not telling you today.

THAT the Twenty-fourth Infantry Regiment joined the Twenty-fifth Infantry Division in Japan on February 1, 1947.

THAT Lieutenant General Benjamin O. Davis Jr. (Retired) U.S. Air Force served as a Lieutenant in the Twenty-fourth Infantry Regiment, 1936-1937, Fort Benning, Georgia.

THAT the four Buffalo Soldier regiments served at one time at the following military posts:

Fort Dodge	Fort Quitman
Fort Gibson	Fort Duncan
Fort Arbuckle	Fort Randall
Fort Richardson	Fort Hall
Fort Griffin	Fort Snelling
Fort Concho	Fort Shaw
Fort Sill	Fort Custer
Fort McKavett	Fort Keogh
Fort Davis	Fort Missoula
Fort Stockton	Fort Meade
Fort Apache	Fort Harker
Fort Grant	Fort Hays
Fort Clark	Fort Arbuckle
Fort Wade	Fort Richardson
Fort Verde	Fort Assinniboine
Fort Buford	Fort Leavenworth

THAT in 1901, a Corporal John E. Green was commissioned a second lieutenant in the Twenty-fifth Infantry Regiment.

THAT during the Johnson County, Wyoming "War" where the cattlemen had an interest in cattle expansion, some settlers resented their plans to move into the areas of Wyoming. Some Buffalo soldiers were involved in maintaining some order in the country.

THAT First Sergeant Gilbert "Hashmark" Johnson was a former member of the Twenty-fifth infantry Regiment prior to his reenlistment in the Marine Corps in World War II.

THAT John T. Pridgen was a member of the Tenth U.S. Cavalry Regiment prior to his enlisted in the Marine Corps in World War II.

THAT the 27th Cavalry Regiment was constituted on November 10, 1942 and assigned to Fort Clark, Texas along with the Ninth Cavalry regiment. The 27th Cavalry was deactivated on February 25, 1943 in Oran Algier.

THAT the late Judge William H. Hastie former civilian aide to the Secretary of War in 1941 and later Governor of the Virgin islands wrote the Secretary of War the following in relation to black units fighting in combat. Hastie wrote: *"I believe Mr. Secretary, that you are entitled to know and the public is entitled to know whether the Ninth and Tenth Cavalry Regiments, and the Twenty-fifth Infantry Regiment have been committed to combat missions for which they have been trained or whether they too have been or are being converted to service units. These four are among the proudest regiments of our army. On the western plains, in the Philippines, with Theodore Roosevelt in Cuba, under Pershing in Mexico, they have been among our first front line troops. Will you inquire, Mr. Secretary, what their missions are today? What we know about the Twenty-fourth Infantry Regiment makes us apprehensive about the other three regiments"*.

THAT in 1878 some young Plains Indians enrolled as students in the Normal Hampton Institute, Hampton, Virginia (now Hampton University). The experiment appeared to be successful and in 1879, an Indian school was started at Carlisle, Pennsylvania. Indian students were also enrolled at Hampton Institute in 1916. (see in Pictorial chapter of this manuscript.)

THAT General Oliver Otis Howard extended an invitation to some Indian leaders to visit Washington, D.C. in 1872 and meet President Grant. The Indian delegation included representatives of the Pimas, Arapahos, Apaches, and White Mountain Apaches. Some of the Indians visited the "College of Deaf Mutes" now known as Galludet University in Washington, D.C. The Indians used their sign language to communicate with the college students.

THAT in 1872 General O.O. Howard housed some Indian leaders during their visit in Washington, D.C. in the dormitories at Howard University.

THAT the chaplains assigned to the black regiments had the responsibility to instruct the troops in reading, writing, arithmetic and provided some spiritual guidance.

THAT some requirements for a cavalryman recruit were "5' 5", not over 150 lbs., 18 years and older and of good character."

THAT during the Indian Campaigns an average cavalrymen's menu could at times consist of "dried beans, coffee, bacon, salt pork, vinegar, sugar, molasses, game beef, corn bread, beans and sweet potatoes."

THAT the horses were trained to lie down, on command. Usually a tap on the horse's left leg, would command the horse to lay down and remain down while the battle was in progress.

THAT during the Indian Campaigns the march formation was:
    Indian scouts
    Advance guard - front, side flanks
    Commander - officers, staff
    Chief of scouts
    Newspaper correspondent
    Surgeon
    Supply wagon, male train
      Ambulance
      Rear guard

THAT the rate of march was "at walk - with horses, 4 miles in 1 hour, 117 yards a minute, trot - 8 miles per hour 235 yards a minute gallop - 16 miles per hour, 470 yards a minute. The best fast movement was an alternating gallop and trot 10 miles per hour, 293 yards. The terrain conditions and weather could affect the rate of movement."

THAT Buffalo Soldiers would execute various tactical formations such as circular movements, construct rifle pits and position an expert marksman on river banks and cliffs.

THAT some tactical movements would move in columns formation, move slowly at a trot, when charging, full gallop, and be in columns of four. The troopers would also use an envelopment, holding force, and when engaging the enemy, move to the right or left and charge in the rear.

THAT a cavalry regiment consisted of three battalions (later squadrons) each containing four companies a total of 12 companies (later troops)

THAT during the Punitive Expedition to Mexico that on April 1, 1916 an advance guard under Major Charles Young confronted 150 Villistas near Agua Caliente. During the exchange of fire, Young ordered his men to execute a tactical maneuver to the left of the Mexicans and was able to route them. Two Villista were killed and Young's soldiers were able to capture a machine gun. When Young's troopers executed a flanking movement against the Mexicans, his Buffalo Soldiers formed abreast in a line of foragers and proceeded down a steep hill. Young signaled the troopers to execute a pistol charge around the right of the Villista's right flank. Tenth Cavalry Soldiers used machine gun support in executing the movement. Young's soldiers were firing over their heads as they forced the Villistas to retreat. It has been said that Major Young and his Troopers introduced a new tactical maneuver of "overhead machine gun fire". It was Major Young's Buffalo Soldiers who came to the rescue of the 13th Cavalry Regiment who was under the command of Major Tompkins.

THAT at Fort Huachuca, Sierra Vista, Arizona there is an eight foot bronze tribute to the Buffalo Soldiers.

THAT when the noted western artist, Frederic Remington wrote an article in 1899 on the Buffalo Soldiers he expressed his real perceptions and impressions of people of color when he wrote:

> *"The Negro troopers sat about their black skin shinning with perspiration".*

Remington also wrote,

"The physique of black soldiers must be admired, great chested, broad shouldered, upstanding fellows with bull necks".

"While the Negro Cavalry men carried on a conversation with their horses. Strange people but yet not half as strange as an Indian in this respect".

The above words of Frederic Remington can be classified as stereotypes.

THAT a distinguished artist displayed his stereotypic personal thoughts about the "Real Buffalo Soldiers". On a visit to African American expositions and festivals, one can find drawings, sketches, painting and figurines, all memorabilia representing the black American military in the west during the Indian Campaigns. The military units were the famed black cavalryman's Ninth and Tenth Cavalry regiments, Twenty-fourth and Twenty-fifth Infantry Regiment affectionately known as the "Buffalo" soldiers. The dealers and vendors are selling reproduction sketches and recent painting of these soldiers. Some memorabilia to include the figures are based on the earlier sketches and drawings of a distinguished artist named Frederic Remington. He made some outstanding sketches and drawings of the black cavalryman. However his reporting of the known inferiorities in education and social customs of the black soldiers did not contribute to the acknowledgement of their outstanding traits and achievements. Remington would often write from his interviews with black troops and use the illiterate dialect or expressions of many black soldiers who were not fortunate to receive a basic elementary education. Remington wrote the following in the *Cosmopolitan Magazine,* February 1897:

"I know why so many of dem battles is victorious" said one trudging darkey to another. *"Dey march de men so hard to get thar, dat dey is too tired to run".*

APPENDIX II

Colonel Charles Young

Charles Young
Colonel, Ninth and Tenth U.S. Cavalry Regiments

Charles Young was born on March 12, 1864, in Mayslick, Kentucky, twelve miles from Maysville in Mason County. He was the son of two former slaves, Armintie Brown and Gabriel Young. His maternal grandparents were Cyrus Bruen and Julia Coleman Bruen, and his paternal grandparents were Peter and Susan Young. His father enlisted on February 13, 1865, as a private in Captain James H. Johnson's Company F, Fifth Regiment of Colored Artillery (Heavy) Volunteers. He served for one year and was discharged on February 12, 1866 near Vicksburg, Mississippi. When he was fourteen months old, Charles Young's parents decided to move across the river to Ripley, Ohio, where his maternal grandmother assisted him in his school work. Charles Young attended grammar school in Ripley, Ohio and graduated with honors.

An advertisement appeared in a Ripley daily newspaper in 1883, stating that examinations for potential military academy cadets would be conducted at Hillsboro, Ohio. The announcement was made by U.S. Congressman Alphonso Hart, from the Twelfth District of Ohio.

Charles Young was teaching school under the supervision of J.T. Whitson, the principal of Ripley High School. Dr. Whitson discussed the newspaper announcement with him and urged him to compete for the West Point appointment. Young was successful, making the second highest score, and in September, 1883, he reported to the military academy as an alternate candidate. In December, 1883, preliminary examinations were given and Young passed twenty-second out of a hundred candidates.

Charles Young arrived at West Point in June, 1884, to formally commence the four-year course. He was turned back to the new fourth class in June, 1885 because of a deficiency in mathematics. He graduated from the military academy on August 31, 1889, number forty-nine in a class of forty-nine members. He was commissioned as a second lieutenant.

After he was appointed to the Tenth U.S. Cavalry Regiment in 1889, the War Department decided to transfer him to the Twenty-fifth U.S. Colored

Infantry Regiment. However, young preferred an assignment to a cavalry regiment and on October 31, 1889, special orders were issued transferring him to the Ninth Cavalry Regiment, his first assignment, at Fort Robinson, Nebraska, on November 28, 1889.

On October 4, 1890, Young reported for duty at Fort Duchesne, Utah. Upon the sudden death of Lieutenant John Hanks Alexander, the second black graduate of West Point, on March 26, 1894, President S. T. Mitchell of Wilberforce University wrote letters to President Cleveland and Senators requesting the appointment of Lieutenant Young as military professor at Wilberforce University. Charles Young arrived at Wilberforce University on May 21, 1894 to assume his new duties. A month later his father Gabriel Young died at the age of fifty-three.

On December 22, 1896, Lieutenant Young was promoted to the rank of first lieutenant and officially transferred to the Seventh Cavalry Regiment (an all-white regiment). The official records did not indicate any specific instances of Young actually serving in garrison with the Seventh Cavalry Regiment (December 22, 1896 to October 1, 1897).

During the Spanish-American War, in 1898, Young was granted a leave of absence from the regular U.S. army to accept appointment in the Ninth Ohio Battalion National Guard as a field grade officer with the rank of major. The Ninth Battalion was assigned to the Second Army Corps, located at Camp Russell A. Alger, near Falls Church, Virginia. In August, 1898, the unit was transferred to Camp George G. Meade, near Middletown, Pennsylvania, and in November, 1898 to Summerville, South Carolina. Young did not see service in Cuba during the Spanish-American War. He experienced the following incident at Camp Algers, Virginia:

A white regiment came in from South Carolina. The men began to call the raw recruits "fresh fish" and some of the new Southern recruits began hollering, "I am not going to be near no damn niggers." After a while, a soldier of that regiment refused to salute Major Charles Young. The camp commandant found it out and one day called this fellow in and also Young. Major Young was told to take his coat off and put it on the chair and then the (white) soldier was told to salute the chair with the coat. Then Young was instructed to put his coat on and the white soldier saluted Young.

After the Spanish-American War, Young returned to his assignment at Wilberforce University. He rejoined his regiment on September 22, 1899, at Fort Duchesne, Utah. While stationed at Fort Duchesne, Charles Young and sixty men were detailed to proceed to the scene of an incident involving Indians and sheep herders to investigate the matter and maintain order. Young was successful in arranging a meeting between the sheep owners and Indians.

The Colored American Magazine of May, 1901 published an article concerning the career of then Lieutenant Benjamin O. Davis, Sr. (later brigadier general). The article mentioned the following concerning Young:

At Fort Duchesne, Utah, where the Third Squadron of the Ninth Cavalry was stationed was Lieutenant Young, at that time the only colored officer in the regular establishment. He became very much interested in Sergeant Major Davis and encouraged him to study and take the examination for lieutenancy. Even the white officers encouraged him to do so and offered him every necessary aid and instruction. Under Lieutenant Young, He applied himself at his severe task for nearly two years studying history, geography, surveying and drill regulations. . .

In February 1901, Charles Young was assigned to the Philippines. He commanded troops at Samar, Blanca Aurora, Daraga, Tobaca, Rosana and San Joaquin from July 1901 to October 6, 1902. He participated in numerous engagements against insurgents on the island of Samar.

On May 20, 1903, Charles Young was appointed acting superintendent of Sequoia and General Grant National Parks, California. He was responsible for the supervision of payroll accounts and also directed the activities of the forest rangers. Colonel Young was transferred on November 2, 1903 and assigned as a troop commander at the Presidio of San Francisco, California. The Visalia Board of Trade, California showed their appreciation of Young's performance of duty as the park's acting superintendent. He was given a citation. On February 12, 1904, a bill was introduced in the House of Representatives to provide for acquiring title to certain patented lands in the Sequoia and General Grant National Parks in the state of California. There was mention of Captain Charles Young in the bill.

As recommended by Captain Young, acting superintendent of said parks, in his report to the Secretary of Interior, dated September twenty-eight, nineteen hundred and three.

While stationed at the Presidio, San Francisco, Captain Young accepted an invitation to address a group of students at Stanford University, (according to the *Daily Palo Alto,* [Stanford, California] December 9, 1903.)

Young reminded the audience that the famed Tuskegee educator, Booker T. Washington's system of race solution would not succeed because when the black man has conquered the industrial trades and is ready for employment, he then confronts the old remark that no Negroes need apply.

On May 13, 1904, Captain Young assumed his duties as military attache to the United States Legation, Port au Prince, Haiti. He was accompanied by his wife, the former Ada Barr. While in Haiti, Young prepared a detailed monograph on the Republic of Haiti (consisting of 284 pages and a map). In July 1905, Young prepared a little handbook of Creole French as spoken in Haiti.

Later Captain Young was accused of sketching fortifications in the interior and gathering information about the Haitian government. Unfortunately, matters did not improve for Young. In 1907 while he was absent on a trip to Cape Haitian, Port de Paix and northern areas some papers were stolen from his quarters. Charles Stephens, a clerk in his employ reputedly sold the papers to the Haitian government for six hundred dollars. The United States government became alarmed about the possible diplomatic complications and the Military Information Division decided on March 26, 1907 that it would be advisable for Captain Charles Young to be relieved of his military attache duty at Port au Prince. Captain Young departed Haiti on April 28, 1907 and reported for duty with the Second Division, General Staff, Washington, on May 7, 1907.

After a temporary assignment in the chief of staff's office, Young rejoined the Ninth Cavalry Regiment as a troop and squadron commander in the Philippine Islands. He assumed command of his troops on August 7, 1908.

In May 1909, Captain Young was assigned to Fort D. A. Russell, Wyoming where he participated in field training exercises and also commanded Troop I and the Third Squadron.

Captain Young reported for duty to the Office of Chief of Staff, United States Army on December 31, 1911. On November 18, 1911, Booker T. Washington wrote Young asking him if he were interested in a position as military attache to Liberia. Young replied that he would be glad to accept the assignment.

On March 8, 1912, prior to his assignment to Liberia, Charles Young was elected as the second honorary member of the Omega Psi Phi fraternity. He also wrote a book, *Military Morale of Nations and Races* published in the same year.

Captain Charles Young arrived in Monrovia, Liberia, on May 2, 1912 to assume the duties of military attache to the American Legation. President D.E. Howard of the Republic of Liberia wrote Major Charles Young a letter on November 23, 1912 requesting him to take an expedition to the relief of a Captain Browne who was surrounded by a group of hostile natives. Young was offered some one hundred men to accompany him on the mission and departed on November 25, 1912 in search of Captain Browne. Young returned to Monrovia on January 15, 1913. He prepared detailed recommendations for improved relations between the government and indigenous tribes for the Liberian government to consider.

In April, 1913, Major Charles Young suffered a very serious attack of malignant malaria (black water fever) and he was granted sick leave and returned to America on May 12, 1913. After recuperating from his illness, he left Ohio to resume his duties in Liberia. He arrived in Liberia for duty on August 11, 1913.

Major Charles Young was officially assigned to the Tenth Cavalry Regiment on September 9, 1915. He reported for duty at Fort Huachuca on or about December 31, 1915. This was Young's first *active duty* assignment with the Tenth Cavalry. On February 22, 1916, Major Charles Young was awarded the Spingarn Medal for his outstanding performances of duty in Liberia. He also served in Mexico with the Tenth Cavalry Regiment as part of the Punitive Expedition. On June 29, 1916 an Army Examining Board found him physically fit and he was promoted lieutenant colonel on September 2, 1916. This promotion was monumental progress. However, there were difficulties ahead. While assigned to the two black cavalry regiments as a junior officer, Young had not experienced much opposition from the other officers who, being white and not directly under the command and supervision of Young, did not complain. However, as Young began to assume command of the regiment and squadron more often, racial prejudice became evident and some officers personally exercised their right to request transfers in order to avoid serving under the black officer.

There was one officer who initiated a campaign to avoid Young's command even before Young reported to the regiment. In May 1915, Lieutenant John Kennard wrote the adjutant general requesting transfer to the Fifteenth Cavalry Regiment, or to any white cavalry regiment serving on the border. A Walter D. Denigre of Manchester, Massachusetts wrote the following to the Secretary of War in December 1915.

"Desire transfer of John Kennard to some other regiment, because he does not wish to serve under Major Charles Young, Tenth Cavalry, who is colored." When John Kennard requested a transfer from the Tenth Cavalry Regiment in May 1915, Lieutenant Colonel Young was still stationed in Liberia. He was assigned to the Tenth Cavalry Regiment on September 9, 1915. However, he did not arrive at Fort Huachuca, Arizona until December, 1915. In August, 1916, John Kennard was transferred to the Seventh Cavalry Regiment. Colonel John Kennard died on December 9, 1959 at New Orleans, Louisiana at the age of seventy. He was eulogized with the following remarks:

He was a man who was representative of the best in the South. In the tradition of the great Confederate Cavalry leaders, he had a wonderful physique, horsemanship, the quick reflexes of an athlete, determination to pursue to the end what he thought was right...

After the Punitive Expedition, Young returned to Fort Huachuca and established a school for enlisted men at the fort. On May 23, 1917, he was directed to proceed to the Presidio of San Francisco, California and report in person to the commanding officer at Letterman General Hospital for observation and treatment. On June 25, 1917, President Woodrow Wilson wrote a personal and private letter to Secretary of War, Newton D. Baker, and stated that a Senator Sharpe Williams of Mississippi was concerned about a first lieutenant in the Tenth U.S. Cavalry, a Southerner who found it not only distasteful but practically impossible to serve under a colored commander." President Wilson asked Secretary Baker if the lieutenant could be transferred and send someone in his place "who would not have equally intense prejudices."

Secretary Baker replied to the president on June 26, 1917 that he had received several letters from senators concerning officers in the Tenth Cavalry who desire not to serve under a black officer. Baker wrote: "The situation is of course, very embarrassing, but I am endeavoring to meet it by using Colonel Young in connection with the training of colored officers for the new army at Des Moines, Iowa. It seems likely that I will be able to tide over the difficulty in that way for at least a while." Secretary Baker also said "There does not seem to be any present likelihood of his early return to the Tenth Cavalry so that the situation may not develop to which you refer." President Wilson wrote Senator John Sharpe Williams of Mississippi that the

"Lieutenant Colonel referred to [Young] will not in fact have command because he is in ill health and likely when he gets better he will be transferred to some other service."

On June 30, 1917, Senator John Sharpe Williams wrote President Wilson that "You seem to have forgotten that it is a Negro regiment as well as a Negro colonel. I send both your letters to me and my letter to you by my secretary so that you can read them (letter) and destroy them, if you choose." President Wilson wrote Secretary Baker again on July 3, 1917 and told him "that Senator Williams said that there is some danger of trouble of a serious nature if this officer is not separated from his present command." On July 9, 1917, Wilson wrote, "I am sorry to bother you [Secretary Baker] about the case [white officer]. . . but the trouble it would seem is not now the fear. . . that he will be put under a Negro officer but that it has got on his nerves that he himself remains an officer in a Negro regiment, and I was wondering whether without violation of the best practices of the department some officer of Northern birth could be substituted for him."

On July 7, 1917, Secretary of War Baker wrote President Wilson: "Prior to your note to me Lieutenant Colonel Young was ordered before a retiring board on the report of the surgeons that he was incapacitated for duty by reason of Bright's disease. Meanwhile, the adjutant general of the State of Ohio has urgently requested his services with the colored command of that state. As soon as the proceedings of the retiring board have been completed and pending final action on them by the War Department, Colonel Young will be directed to report to the adjutant general of the state of Ohio for the above duty. This, I think, will remove the cause of trouble so far as I now understand it. The Colonel [of the white officer] is a white officer, as are also the other officers of the regiment." The War Department issued orders on July 10, 1917 assigning Colonel Young to the adjutant general of Ohio for duty. Colonel Young was later retired from active service and remained in the area of Wilberforce, Ohio.

On June 6, 1918, in an attempt to prove his physical fitness for active duty Colonel Young rode on horseback and walked one quarter of the way from Ohio to Washington, four hundred and ninety-seven miles and unattended. Young arrived in Washington on June 22, 1918 having rested only one day. The trip lasted sixteen days. Colonel Young was recalled to active service in 1918, though the war was near its end. He was assigned to Camp Grant, Illinois, and he returned to Monrovia, Liberia as military attache in 1919.

On November 15, 1921, Colonel Young departed on the Spanish steamer, *Catalina* for Fernando Po and Nigeria. Colonel Young was admitted on December 25, 1921 to Gray's Hospital, Lagos, Nigeria suffering from nephritis, medically reported as "acute exacerbation of an old-standing complaint." He was attended daily by a European physician, Dr. Aitken. He died on January 8, 1922, and was given a military funeral in Lagos, Nigeria and buried in an English cemetery. His body was brought to the United States in 1923 and on June 1, 1923, America's third black West Point graduate was given a hero's burial in Arlington National Cemetery.

Charles Young and Troop M, Ninth Cavalry 1903
National Park Duty, 1903

Captain Charles Young was assigned as the acting superintendent of Sequoia and General Grant National Parks, California on May 20, 1903. On June 3, 1903, Young wrote the Secretary of Interior a letter requesting instructions for his new assignment. He also informed the secretary that he was making preparations for a road improvement project and that it was his intention to follow his predecessors' plans and also initiate measures to prevent trespassing. Young more specifically explained the importance of the road project to the Secretary of Interior in a telegram, dated June 4, 1903. He stated:

> *Request permission to begin work on the giant forest road immediately in Sequoia National Park to be paid from the appropriation fiscal year 1904. Laborers are on the ground now. Many hundreds of dollars will be paid by the government by not waiting until the dry season July.*

Captain Young was responsible for the supervision of the payroll accounts and also directed the activities of the forest rangers. He was concerned about the method of payment for the laborers and was interested in expediting payments in order to enable projects to continue. In a letter to the Secretary of Interior dated June 10, 1903, Young requested permission for the officers in charge of the work projects to be permitted to vouch for payment of laborers debts while they are working in the mountains. He felt the accounting system and assurance of contractors paying the laborers could be improved.

Young had been instructed by the Secretary of Interior to submit a brief monthly report of his activities. On June 30, 1903, he wrote the Secretary of Interior the following:

I have the honor to state that during the past month work has been pushed with all diligence in the Giant Forest with most successful results. I am bending every effort toward completing this road into the forest this year and by economizing in the appropriation as much as possible. I thing it may be sufficient for this period. I am also working with respect to the private land claims and hope later to report satisfactorily upon the solution of this the greatest concern relative to the parks. The avenues of entrance to the parks are well guarded, my detachments and forest rangers are efficiently cooperating toward the protection of game and prevention of grazing on public land. The work on the road in General Grant Park begins the first week in July and the Fresno County authorities have kindly consented to aid in connecting with them.

Captain Young's payroll for civilian employees for the month of June consisted of sixty workers; blacksmith, gang foreman, stump bluster, powderman, cook, driller, timber faller, flow holder and forty laborers. The total expenses incurred were $3,707.62.

Young and his detachment were stationed at the National parks during a period when the government was attempting to purchase additional land for the park ($25 an acre). Correspondence of a George W. Stewart, dated January 17, 1904 refers to a project that was completed by several of the Ninth Cavalry soldiers and a park ranger. The letter said:

Last summer Park Ranger Ernest Britten assisted by Corporals Mosby and Smith of Troop 1, Ninth Cavalry, made a count of Redwood trees on NW + of Sec. 6, T.16 S.R. 30E, M D. being and eighty acre tract within the giant forest and the result showed 185 trees.

Captain Charles Young's duties also involved negotiating with the local county authorities, visits to Mount Whitney Reservation and the development of improvement plans. On August 29, 1903, Young submitted the following plan to the Secretary of Interior:

1. Completion of the road of last year to S.W. corner of park - the cost of powder, tools and labor estimated at $1500 (This is the most important work in connection with this part for reasons stated hereafter).

2. Cleaning up rubbish and dead and down timber near the big trees and constructing firebricks on the west and north sides of Park to

prevent spread of fire to the part from the engine and log camps of the Sawyer Lumber Company - estimated cost. $100.

3. Constructing fence about General Grant Tree and getting out lumber for a small stable for protection of the animals of the forest rangers during the winter, cost about $100.

4. Continuing as far as the remainder of the appropriations will permit of the main road toward King's Canyon which has been for years an important point of tourist travel because of its scenic value.

I beg that this plan be approved and the expenditure for 1st and 2nd parts thereof be allowed, as they have been about completed by this time as Part 1 is the only safe wagon travel into the park.

Captain Young demonstrated his modesty on many occasions during his military career. In writing to the Secretary of Interior about the Parks progress he wrote:

The road has been completed into Giant Forest and around to Moro Rock, water has been placed upon the road for convenience of the traveling public and trails are being repaired. I submit that more work has been done in any two years previous to this. I claim no special credits for this (my emphasis) as it is largely due to the department's permission to allow work to begin early in the season when the ground was moist and where good men were available.

Young was always concerned about his subordinates' welfare and openly expressions of his feelings on this subject were made. He mentioned to the Secretary of Interior in his correspondence of August 29, 1903, the need for authority to adjust the vouchers and to prevent delay in the payment of work accomplished during the month of August. He indicated his interest in the employee's welfare when he stated:

An additional reason for this last request will be found in the fact that these men are poor and any delay in their accounts will work great hardships.

In September, 1903, a captain assigned to the Ninth Cavalry who was senior to Captain Young as far as date of rank was concerned was instructed to report to his Troop M, which was assigned at the Sequoia and General Grant National Park. Major General Arthur MacArthur, commanding general, headquarters, Department of California, selected Captain Lester W.

Cornish, Ninth Cavalry to replace Captain Charles Young as Acting Superintendent, Sequoia and General Grant National Park, California.

The Secretary of Interior had requested on September 18, 1903 information relating to a list of patented lands in the Parks. Young was instructed to investigate this matter and submit a report. Captain Young forwarded a reply to the Secretary of Interior on September 28, 1903. Young wrote the following:

I have carefully examined most of the lands, duly considered the offers here with (he forwarded agreements that claimants had concurred with) in connection with the forest rangers, the Register of the land office and some of the best businessmen of this section, all whom are highly interested in the welfare of the park and in the purchase by the U.S. at a reasonable figure of these lands and we all think and submit that the prices asked in the agreement are reasonable and in many cases low. Out of 18 owners in Sequoia National Park the agreements of 13 are herewith submitted, the other 5 with one exception are small owners whom it has not been possible to reach yet, but by estimate and comparison with the other agreements it is thought that their claims will not exceed $16,000, while the other claims aggregate $70,734, these with an offer of 160 acres in General Grant Park for $1600 (the only private claim in the Park) will bring the entire claims within $73,000, which amount is but a trifling sum in comparison with the benefits accruing to the government in securing for the nation at least 20,000 Giant Sequoias with innumerable young trees of this species and other timbered and meadow lands all amounting to 3,877 acres. The price asked averages about $18 per acre. As the agreements for the most important of these lands, that is those in comprising the "Giant Forest," only remain in force for a year, I earnestly urge that immediate steps to be taken to effect their purchase.

The Acting Secretary of War informed the Secretary of Interior on September 30, 1903 that the commanding general, Department of California had selected Captain Lester W. Cornish, Ninth Cavalry, to replace Captain Charles Young as the Parks' acting Superintendent.

When Captain Cornish assumed the duties of acting Superintendent of the Sequoia and General Grant National parks he wrote the Secretary of Interior the following:

In reference to the report of Captain Charles Young concerning the purchase of private lands in the Sequoia National park, I have the honor to heartily endorse each and every recommendation made by him in his report. Captain Young is to be congratulated upon the remarkable success he has obtained, and I earnestly recommend that his suggestions be carried out in every respect.

Captain Young was always willing to support his subordinates and to honestly admit and defend his actions if they were challenged by his superiors. The Secretary of Interior had inquired about ten days of absence with pay that was granted a forest ranger by Captain Young. Young responded to the inquiry by explaining his actions and also offered to repay the government the money paid the ranger for ten days absence. Captain Young wrote:

As was stated in my annual report, this ranger (Ranger L.L. Davis) volunteered to superintend the work of blasting this season on the Giant Forest Road, thereby saving many hundreds of dollars to the government by his good sense, good judgment, and hard work.

At the end of the season because of the ill effects the close contact and long use of the dynamite had worked upon his system and general health. I ordered him away from duty for ten days to rest. He went unwillingly, but I felt that he was too valuable a man in his place here to sacrifice when the rest could put him in his usual form again. So far as he was concerned he was on duty obeying my order as Acting Superintendent of the Parks. I therefore request that my action be approved by the department and if the exigencies of ranger service will not permit him to have those ten days so richly deserved by him, <u>I shall be glad to refund the money paid him by the department for those days</u>. (My emphasis).

Captain Charles Young had performed an outstanding service to the Interior Department within a brief period of almost six months. His adeptness of administration and ability to supervise civilians as well as military personnel contributed to his success. The outstanding performances of duty by Captain Charles Young and his responsible and dedicated men of Troop M is another true fact that those Real Buffalo Soldiers had the abilities and personal self concept to achieve excellence in their assigned duties.

COLONEL YOUNG AND THE Ninth CAVALRY ON THE UTAH FRONTIER

In 1898 attempts were made by the federal government to negotiate a treaty with the Uintah Indians who were settled near Fort Duchesne. Many

Indians did not show any interest in occupying the new allotted land, some 70,000 acres. The soldiers at Fort Duchesne had the responsibility of searching for discontented or renegade Indians and also initiating efforts to maintain peace between the Indians and the white settlers. It was necessary for the troopers to certify weekly beef rations for the Indians. Each Indian received seven pounds of beef per week. The black cavalry troopers were present at Fort Duchesne, and assisted in maintaining peace and order on the frontier.

The Indian situation began to create several problems for the federal authorities in July, 1899. There were three main tribes located near Fort Duchesne. They were the Uintahs, White Rivers, and Uncompahgres. An Indian agent stationed 14 miles from the fort was responsible for negotiations with these tribes. An Indian Chief, called Sowawick of the White River Indians was displeased with the general situation and wanted to move to Colorado where he felt his people should be located. Chief Sowawick demanded that if he must remain at the reservation, that he wanted the reservation divided. He felt the Uintahs and Uncompahgres should be separated from his tribe. The White River Indian Chief also requested that all white men be driven out of an area known as Ashley Valley. He also did not appreciate farming and did not want the soldiers from Fort Duchesne observing their movements at all times.

The federal Indian agents informed Chief Sowawick that he must remain on the reservation and that if he did leave he would be punished. He was not granted any other requests.

The other Indian chiefs located near Fort Duchesne appeared to be satisfied. The Chief of the Uintahs, Tabby, was concerned about an Indian agent that allowed some sheep men to use part of his reservation without the Indians' permission. An investigation of this matter revealed that the Indian agent was receiving rent money from the sheep men with intentions of distributing the money among the Indians. The agent stated that he had official approval for the transaction.

On November 8, 1899, a verbal report was received by Captain John F. Guilfolge, Ninth Cavalry, commander of the post that a sheep herder had been seriously wounded and one Indian killed. The wounded white man was brought to the Post and treated by the surgeon. The incident occurred near Rock Creek, approximately sixty miles from the post. The Indians had surrounded the sheep herders but some were able to escape.

The wounded sheep herder related his version of the incident to Captain Guilfolge. He reported that:

"Two Indians came to the sheep camp and one of them made a threatening gesture with his riding whip toward his (wounded man) companion herder whereupon this man attacked the Indian with a tent pole. The Indian, being thus attacked, jumped from his horse, and seizing an ax lying near by, prepared to use it, the herder almost immediately jumped upon the Indian getting the ax, opened fire upon him with his revolver, firing from it five shots. Having emptied his revolver, as he supposed, he threw it away from him and resumed with his tent pole. The Indian picked up the weapon and finding one load remaining, fired at the herder who jumping behind his partner caused that individual to receive the bullet in his arm. The herder with the tent pole then again attacked the Indian beating him over the head with it to insensibility.

Captain Guilfolge made the decision to retain the sheep herder at the post. He also dispatched Lieutenant Charles Young and sixty men with a pack rain and rations to the scene of the incident to investigate the matter and maintain order.

The Adjutant General in Washington was informed of this incident on November 10, 1899.

Lieutenant Charles Young was briefed concerning the incident. He was told that a white man named Olsen had killed an Indian known as "Mountain Sheep" and that a Mexican sheep herder was wounded. Young's specific instructions were as follows:

"You will proceed as early in the morning as possible to the scene of this disturbance with a detachment of sixty men of Troop I, pack train with 10 days rations, 100 rounds of ammunition to a man and as much extras as you may deem advisable. Preserve the peace between the herders and the Indians, make thorough investigation of the matter and report your actions to me; return to the post as soon as possible. Take precautions to make those arrangements that should obtain marching in an enemy's country."

Young and his troopers departed the post at 11:30 a.m. Thursday, November 9, 1899. The detachment reached an Indian settlement on the Duchesne River around 7:00 p.m. after traveling an estimated 40 miles. They camped for the night and were some 18 miles from Mountain Sheep's ranch. Lieutenant Young, using the services of an interpreter, inquired among the neighboring Indians about the cause of the altercation. He learned that an argument had ensued after a discussion among the Indians and sheep herders on grazing land. He was also told that Mountain Sheep was critically wounded and all of the male Indians of the immediate settlement were at Mountain Sheep's ranch.

On Friday, November 10, 1899, Young and his detachment proceeded 14 miles toward a camp about 4 miles from Mountain Sheep's ranch. Lieutenant Young described very dramatically and in detail how he and his men had approached the ranch. He stated as follows:

> *As grass, wood and water were here (4 miles from Mountain Sheep's ranch) in abundance, I ordered the camp to be pitched permanently, while I took the guide and 18 men and went on to Mountain Sheep's place. On the way we met several herds of sheep but their herders had all fled, so the Indians informed me. Runners had doubtless given the Indians notice of our coming as they had taken every precaution for defense in event of attack, being deployed behind rocks and on the hills round about the tents of the wounded Indian. Upon arrival, I dismounted the detachment to let it rest while I took the interpreter and went into the tent of Mountain Sheep to hear his story relative to the disturbance and to examine his wounds which I found serious enough. One pistol shot had penetrated his jaw coming out the back of his neck; another had gone into his left breast, coming out in his back. Besides this, his head was covered with wounds, and his arm he feared was broken; this we found on the following day was only a sprain.*

Lieutenant Young believed that the Indian's sworn statement was true and reported that the incident was caused by the renting of grazing land to sheep owner without the Indian's consent. There were also allegations of Mountain Sheep's water supply being contaminated by the shepherds. It appeared that Mountain Sheep had requested the herder to remove his sheep some 500 yards below the water source. The herder refused and when Mountain Sheep continued to protest, he was shot by the herder, Olsen.

Charles Young said that he had found the Indians willing to permit the troops to assist them in settling their grievances, but they were in agreement that the sheep herds must be removed. Later Young met some of the sheep herders near his camp and they agreed to remove their herd if the troops would provide them protection. In order to prevent any further problems, Young agreed to remain for three days to protect the herders.

Lieutenant Young was successful in arranging a meeting between the sheep owners and the Indians. The group met at Mountain Sheep's ranch and all parties agreed that a peaceful understanding could prevail if the sheep herds were removed from the Indian reservation and that Mountain Sheep's assistant should be punished accordingly. The detachment patrolled the area daily using a squad of 15 to 18 men. Young used this measure in order to prevent any further trouble between the herders and Indians.

The detachment of sixty men and their capable leader had accomplished their mission in a most outstanding manner. Lieutenant Charles Young exhibited his abilities of leadership, command and initiative during this important frontier mission. He remarked in his official report how it allowed his men opportunities to receive valuable training during the march and camping. Young wrote the following in his report:

In addition to the main duty imposed upon the detachment, its march was so arranged that enroute to the (sic place of disturbance the covering force was relieved and another formed from a different part of the detachment with another chief to command it; thus giving this valuable practice to all the non-commissioned officers and placing all the men either in the planning or the advance patrols of the vanguard under actual conditions. No time was lost in these changes as they were made at the hourly halts. At different places on the march non-commissioned officers were questioned as to the disposition they would make to meet an attack, and an immediate answer, whether right or wrong, exacted. On the return march, field notes were taken by all the non-commissioned officers, distances calculated from rate of travel of their horses and the bearings were taken by them to prismatic compasses. I propose later to teach them how to plot these notes.

Young and his detachment were also observant of the country side and its terrain features during the course of their marches. Lieutenant Young showed his earnest appreciation of his men by commenting in his report that the behavior of his men was good and that there were no complaints about the hardships and duties that confronted the men during the challenging mission.

Captain Guilfolge forwarded a report on November 23, 1899 to the Adjutant General Department of the Colorado informing him that the incident had been settled. He also remarked about the efficient manner in which Lieutenant Young had executed his orders related to the incident.

The commanding general, Department of Colorado wrote the Adjutant General, Washington about the conclusion of the Indian and sheep herders conflict. He commended Lieutenant Young in his letter. General Merriam wrote:

> By his (Lieutenant Young) prompt action and good judgment, as well as that of his post commander, further bloodshed has probably been averted for the present.

It is interesting to note that General Merriam referred to in his report, the incident of August 8, 1899, which was a warning of possible Indian unrest concerning grazing land and sheep owners. The General also expressed his personal feelings of the Indians when he referred to them as "savages".

The incident between the white sheep herder and Indians from the Uintah reservation could have developed into a serious situation. However, the negotiations, investigation and efforts to preserve the peace were essentially accomplished by sixty men under the leadership of a lone black regular army officer on active duty Lieutenant Charles Young.

Buffalo Soldiers Wife's Concern

The woman beside the renowned Buffalo Soldier and race hero, deserves some consideration and laudation for her continuing assistance and love for her husband. During Colonel Young's ordeals and triumphs, Mrs. Ada Barr Young was there assuming the role of a woman not behind but beside a great man. One of her behind the scene actions that demonstrates a loyal wife's interest in her husband's future is depicted in a letter Mrs. Young wrote to General John J. Pershing on 12 October 1921 while she was residing at 9 Rue Jasuvin XVI, Paris, France. Because of this unusual action and the heart-felt emotion of her husband's problems and her concern for the family, the entire letter is being quoted. Even though General Pershing did not respond positively to her request, the importance of her letter probably touched him in some personal manner because 58 years later the original correspondence could be found in General John J. Pershing's personal papers.

My dear General Pershing:

I had intended asking to see you for just a few minutes, but knowing how taken up your time is and your visit so short, every minute must count, so I am just going to write and you can read it at your leisure. I will consider it an honor to have an answer from you, writing one with absolute frankness, just what you feel and think is this matter. If I am asking too much you will know its just an anxious wife and mother. One who is just as proud of your success and achievement as anyone who has watched your career with interest for I know members of your family. It is with regards to my husband and I am writing. He served with you in Mexico and you _know the man_ his long years of service, loyalty and worth.

His ordered retirement from orders for his reinstatement were pigeoned holed. You have the chance of doing a great-deal of good in helping your black brother. We all know how you have aided the black soldier and the colored people as a whole in the USA. Would so appreciated Col's promotion as they know he is the only one who could get it and now that there are none to follow I can't see why the point couldn't be if Congress stretched, do you? Of course, I know the act of Congress (and) etc., but other offices have been created and officers made, so it could happen, you see. When I tell you in Africa he had to live, keep house, the children, too and I have to live because of the climate we can't be together. I tried it, you know, and his mother provided for, it means just three households and retired pay won't do it. You have an idea what living is and as modestly as I live. If he could be retired as General the pay would be enough, or if with the advanced rank, and a small foreign post somewhere in Europe I suppose one would have to be created. You would know better than I do, would mean so much to him. Now perhaps you think me presumptuous, but I do not mean it so and will take it kindly if you write me just what you feel and think about it and will do about it.

I just ask you not to let him know it was me that prompted you to do something in his behalf. I would rather he feel it was you. But I just felt it was no harm to write and ask you because with all you have in mind, you might not think. In my California home an old man used to say, "What's worth having is worth asking for". Wishing that all good may attend your success.

Very respectfully, Ada M. Young

APPENDIX III

The Secretary of Interior, Albert B. Fall's letter

The Secretary of War, John W. Weeks, replied to Secretary Albert B. Fall's letter on October 6, 1922, as follows:

Honorable Albert B. Fall, Secretary of the Interior, Washington, D.C.

My dear Mr. Secretary: I acknowledge receipt of your letter of September Ninth, enclosing a copy of the letter written by you to Senator Wadsworth with reference to the bill for the relief of Mr. H. O. Flipper, an employee of the Interior Department, who, as an officer of the Tenth Cavalry, was dismissed from the military service on June 30, 1882.

I find that before a Court-Martial which convened in November, 1881, Second Lieutenant Henry O. Flipper, Tenth Cavalry, was tried under two charges, the first being "Embezzlement, in violation of the Sixtieth Article of War" of the sum of $3,791.77 public funds; the second, "Conduct unbecoming an officer and a gentleman," with five specifications which respectively were, false official statements made to his commanding officer on four different occasions to the effect that he had transmitted to the Chief Commissary at San Antonio, Texas, on July 9, 1881, $3,791.77, which funds "were not so in transit but had been retained by him or applied to his own use or benefit." The fifth specification under this charge was that when required officially to make an exhibit of the public funds in his personal possession he showed to his commanding officer s part of the funds for which he was accountable a check for $1,440.43 "which check as fraudulent and intended to deceive the said commanding officer, as he, Lieutenant Flipper, neither had nor never had, personal funds in said bank, and had no authority to draw said check."

The action of the court was to acquit Lieutenant Flipper on the charge of embezzlement, which was undoubtedly due to some technical construction of the term embezzlement, for the court proceeded to find him guilty of all the specifications under the other charge, which involved the retention by him of public funds to be applied to his own use of benefit. Under the charge of conduct unbecoming an officer and a gentleman, supported by the five specifications which I have explained, the finding of the court was guilty and the sentence was dismissal, which was carried into effect June 30, 1882.

It is proposed to restore this man to the grade, rank and status in his arm to which he would have attained had he remained continuously in the service until the date of the approval of this Act, which would make him the senior colonel in the army, and place him on the retired list. The best that this government can do for the most valuable officer in the service, the one with the longest service, of the most gallant conduct, and of the most unblemished and irreproachable record is to place him on the retired list at the end of his continuous and faithful service. Officers are only placed on the retired list when stricken with permanent illness or disabled by wounds, or after at least thirty years of service. This reward your request would give to an officer educated by the government, who within five years of his graduation was disgracefully dismissed from the service, and in the intervening forty years has rendered no service whatever to the army. I believe if you will consider the matter, you will agree that I can not with any propriety concur in this proposition. To place a man of this record on the retired list of the army would be a reflection on every honorable officer on it and would give him, unearned, all that comes to other officers after the devoted service of a life-time.

Sincerely yours, (Signed) "John W. Weeks, Secretary of War.

Evidently Henry O. Flipper had an opportunity to refute some of the statements that were written by Secretary Weeks to Secretary Fall. The following is excerpted from Flipper's remarks. He wrote:

The statement "which was undoubtedly due to some technical construction of the term embezzlement" IS GRATUITOUS.

The record shows that the charge of embezzlement under the Sixty article of War was abandoned and an effort made to convict for constructive embezzlement under sections 5488, 5489, 5490, 5491, 5492 and 5493 of the Revised Statutes. Under neither of which sections was there charge indictment, or arraignment for trial, hence the ACQUITTAL, See original record pp. 153 to 169, pp. 7 to pp. 10 of Argument of Counsel for Defense.

As to the check, the commanding officer himself testified:

"The only time I can swear positively to observing the check was on the 8th day of July, when I spoke to him in reference to it." Record, Certified Copy, pp. 93, 94.

"Q. - Then, on Twenty-sixth or 27th of June, on the occasion which you speak of, he had ALL THE FUNDS with which he was responsible?'

"A. - Undoubtedly. I did not inspect him on the Twenty-sixth of June as the papers show, but I did inspect him on the 8th of July up to that date and he had ALL THE MONEY that he was accountable for. It was public money and I considered it correct."

WHERE IS THE GUILT?

The commanding officer had previously testified the check had been submitted to him sometime "in May and thereafter weekly until the 6th of July," Recorded, Certified Copy, Part 1, p. 46.

As a matter of fact the check was first submitted in May, was made and submitted to commanding officer by his orders to make such check and forward funds to some bank as personal funds. The check was never forwarded because of his order to hold it till further orders.

The Secretary of War is in error as to retirements. The most inconspicuous officer of the lowest or any other grade in the army is entitled to retirement after 40 years' service at this own request, as well as for disability incurred in the service. See Revised Statutes, sections 1243 et seq.

As to "reflection on every honorable officer," did the restoral of Captain George A. Armes, Paymaster Major Reese, Judge Advocate General D. G. Swain and other mentioned in my Brief as precedents impose any less refection on these gentlemen? Henry O. Flipper.

BIBLIOGRAPHY

BIBLIOGRAPHY

Primary Sources

Manuscript Collection

Manuscript Division, Library of Congress, Washington, D.C.,

Newton D. Baker Papers
Christian Fleetwood Papers
John J. Pershing Papers

Moorland Spingarn Research Center, Howard University Washington, D.C.

Mary O.H. Williams Collection

Soper Library, Manuscripts, Morgan State University, Baltimore, Maryland

Emmett J. Scott Papers Collection

The Ohio Historical Society

Asa S. Bushnell Papers
George R. Meyers Papers

National Archives, Washington, D.C.

Military Pension Files RG 15 of the Adjutant Generals' office

Military Intelligence Division Papers RG 165

Camp McGrath, Batangas, Philippines Island *Post Returns* RG. 98, 1908-1909

Fort D.A. Russell, Wyoming *Post Returns* RG 98, 1909-1911

Fort Duchesne, Utah *Post Return* RG 98, 1890, 1891, 1892, 1894-1901

Ninth Cavalry *Regimental Returns* RG 94, 1889, 1890, 1894, 1895

Punitive Expedition, Mexico, 1916-1917 RG 395

Records Of The National Park Service, RG 79, 1903

Tenth Cavalry Regiment, *Regimental Returns* RG 94, 1915, 1916

Newspapers

Afro American (Washington, D.C.)
New York Times
Washington Post

Periodicals and Magazines

Journals

Cavalry Journal
Colored American
Cosmopolitan
Journal Negro History
Negro History Bulletin
Utah Historical Quarterly

Magazines

Crisis
Ebony
Jet

Interviews

Matthews, Mark, First Sergeant, "Conversation with Mark Matthews" Interviewed by Robert E. Greene, July, 1994, Washington, D.C.

Personal Notes

Unpublished and published manuscripts, class lectures and photographs from the library of Robert Ewell Greene.

Secondary Sources

Books

Andrist, Ralph K. *Long Death The Last Days of The Plains Indian.* New York: MacMillan Publishing Co., 1993.

Armes, George A. *Ups and Downs of an Army Officer.* Washington: 1900.

Barrett, S.E. *Geronimo His Own Story.* New York: Ballantine Books Inc. 1972.

Basco, Keith H. ed. *Western Apache Raidings and Warfare.* Tuscon: The University of Arizona Press, 1993.

Beyer, W. F. and Keydel, O.F. *Acts of Bravery Deeds of Extraordinary American Heroism.* Detroit: Keydel Co., 1903

Bigalow, John. *Reminiscences of the Santiago Campaigns.* New York: Harpers Bros. 1899.

Bigglestone, William E. *They Stopped In Oberlin, Black Residents and Visitors of the Nineteenth Century.* Arizona: Innovation Group Inc., 1981.

Blockson, Augustus P., et al. *Affray at Brownsville, Texas.* 1900. Washington, D.C.: U.S. Government Printing Office, 1906.

Brown, Dee. *Bury My Heart at Wounded Knee.* New York: Henry Holt and Co., 1991.

Brown, Dee and Schmitt, Martin F. *Fighting Indians of the West.* New York: Ballatine Books, 1974.

Browne, Frederick W. *My Services in the U.S. Colored Cavalry.* Cincinnati: 1908.

Burt, Olive W. *Negroes In The Early West.* New York: Julian Messner, 1969.

Cantor, George. *North American Indian Land Marks a Travelers Guide.* Detroit: Visible Ink Press, 1993.

Cody, William F. *The Life of William F. Cody (Buffalo Bill).* New York: Indian Head Books, 1991.

Connell, Evan S. *Son of the Morning Star Custer and the Little Bighorn.* New York: Promontary Press, 1993.

Coston, Hilary W. *The Spanish American War Volunteers.* 1899.

Cox, Clinton. *The Forgotten Heroes, The Story of The Buffalo Soldiers.* New York: Scholastic, Inc, 1993.

Custer, George A. General. *My Life On The Plains.* Oklahoma: 1876.

Davis, Britton. *The Truth about Geronimo.* New Haven: 1929.

Durham, Philip and Jones, Everett L. *The Negro Cowboys.* Lincoln, Nebraska: University of Nebraska Press, 1965.

Downey, Fairfax. *Buffalo Soldiers in the Indian Wars.* New York: McGraw-Hill, 1969.

Editors, Time Life. *The Wild West.* Time Life Books, 1993.

Emilio, Luis F. *The Assault On Fort Wagner.* Boston: Rand Avery Co., 1887.

Erdoes, Richard. *The Sun Dance People, The Plains Indians Their Past and Present.* New York: Alfred A. Knopf, 1972.

Faulk, Odie B. *The Geronimo Campaign.* New York: Oxford University Press, 1969.

Flipper Henry O. *The Colored Cadet at West Point.* New Hampshire: Ayer Co. Publisher Inc., 1986.

Fowler, Arlen L. *The Black Infantry In the West 1869-1891.* Wesport: Connecticut: Greenwood Publishing Co. 1971.

Franklin, John Hope. *George Washington Williams. A Biography.* Chicago: The University of Chicago Press, 1985.

Garland I., et. al. *The University of Life or Practical Self Educator.* Nashville, Tennessee: Southwestern Co., 1900.

Glass, Edward. *History of The Tenth Cavalry 1866-1821.* Arizona, 1921.

Grafton, John. *The American West In The Nineteenth Century.* New York: Dover Publications Inc., 1992

Greene Robert E. *A Biographical Study of The 54th Massachusetts Regiment, Swamp Angels.* Fort Washington, Maryland: BoMark/Greene Publishing Group, 1990.

Greene, Robert E. *Black Courage 1775-1783.* Washington, D.C.: National Society of the Daughters of The American Revolution, 1984.

Greene, Robert E. *Black Defenders of America 1775-1973.* Chicago: Johnson Publishing Co, 1974.

Greene, Robert E. *Black Defenders of The Persian Gulf War, Desert Shield Desert Storm.* Fort Washington, Maryland: R.E. Greene Publishers, 1991.

Greene, Robert E. *Charles Young Soldier and Diplomat.* Washington, D.C: R.E. Greene Publishers, 1985.

Greene, Robert E. *They Did Not Tell Me True Facts About African Americans In the African and American Experience.* Fort Washington, Maryland: R.E. Greene Publisher 1992.

Greene, Robert E. *They Rest Among The Known.* Washington, D.C.: Yancy Graphics, 1981.

Greene, Robert E. *True Tales For Children Young Adults and Adults.* Washington D.C.: R.E. Greene Publisher, 1987.

Hare-Cuney, Maude. *Negro Musicians and Their Music.* Washington, D.C. The Associated Publisher, 1936.

Howard, Oliver O. *My Life and Experiences Among Our Hostile Indians.* Connecticut: 1907

Johnson, Barry C. *Flipper Dismissal: The Ruin of Lt. Henry O. Flipper. First Colored Graduate of West Point.* London: 1980.

Johnson, Charles Jr. *African American Soldiers In The National Guard.* Connecticut: Greenewood Press, 1992.

Johnston, Terry C. *Dying Thunder The Fight At Adobe Walls and The Battle of Palo Duro Canyon, 1874-1875.* New York: St. Martins Press, 1992.

Johnston, Terry C. *Shadow Riders, The Southern Plains Uprising, 1873.* New York: St. Martins Press, 1991.

Johnston, Terry C. The Stalkers, The Battle of Beecher Island, 1868. New York: St. Martins Press, 1990.

Katz, William Koren. *The Black Presence In The West.* New York: Anchor Press, 1971.

Knowles, Thomas W. and Lansdale, Joe R. *The West That Was.* New York: Wings Books, 1993.

Lamb, Daniel Smith. *Howard University Medical Department.* Washington, D.C.: Beresford Printers, 1900.

Leckie, William H. *The Buffalo Soldiers A Narrative of The Negro Cavalry in the West.* Oklahoma: University of Oklahoma Press, 1967.

Lee, Irvin H. *Negro Medal of Honor Men.* New York: Dodd, Mead and Company, 1967.

Lee, Ulysees. *The Employment of Negro Troops.* Washington, D.C.: U.S. Government, 1966.

Lindenmeyer, Otto. *Black and Brave. The Black Soldier In America.* New York: McGraw-Hill, 1969.

Longstreet, Stephen. *Indian Wars of The Great Plains.* New York: Indianhead Books, 1970.

Lynk, M. V. *The Black Troopers.* Jackson Miss: M.V. Lynk Publishing Co. 1899.

MacDonald, F. *Insights Plains Indian.* New York: Barrons Educational Series, 1993.

Malone, John. *The Native American History Quiz Book.* New York: William Morrow, 1994.

Mancini, Richard. *American Legends of The Wild West.* Pennsylvania: Running Book Publishers, 1992.

Matloff, Maurice ed. *Army Historical Series, American Military History.* Washington, D.C.: U.S. Government, 1969.

McIntyne, Irwin W. *Colored Soldiers.* Macon Georgia: Burke, 1923.

Miller, William G. *The Twenty-fourth Infantry Past and Present.* N.P. 1923.

Nankivell, John H. *A History of the Twenty-fifth Regiment United States Infantry.* Denver: Smith Brooks Printing Co. 1927.

Nye, Wilbur S. *Plain Indian Raiders.* Oklahoma: The University of Oklahoma Press, 1968.

Pipkin, J. J. *The Negro In Revelation in History and in Citizenship.* St. Louis, MO: A.D. Thompson Publishing Co, 1902.

Raphael, Ralph B. *The Book of American Indians.* New York: Arco Publishing Co., 1973.

Reedstrom, E. Lisle. *Apache Wars An Illustrated Battle History.* New York: Sterling Publishing Co., Inc., 1992.

Roberts, David. *Once They Moved Like The Wind, Cochise, Geronomo and The Apache Wars.* New York: Simon and Schuster, 1993.

Rodenbough, Theodore and Haskin, William L. (Eds.). *The Army of The United States.* New York: Maynard and Merrill Co., 1896.

Scott, Emmett J. *The American Negro In The World War.* New York: Arno Press and The New York Times, 1969.

Schoenfield, Seymour J. *The Negro in the Armed Forces.* Washington, D.C.: Associated Publishers, 1945.

Simmons, William. *Men of Mark.* George M. Rewell & Co., 1887.

Sun Bear. *Buffalo Hunts.* California: Naturegraph Publishers, 1970.

Stewart, Theophilus G. *The Colored Regulars in the United States Army.* Philadelphia: AME Book Concern, 1904.

Taylor, Colin *What Do We Know About Plains Indians?* New York: Peter Bedrick Books, 1993

Utley, Robert M. *Billy The Kid A short and Violent Life.* Nebraska: University of Nebraska Press, 1989.
Utley, Robert M. *Fort Davis National Historic Site, Texas.* Washington, D.C. National Park Service Historical Handbook Series. No. 38, 1965.
Utley, Robert M and Washburn, Wilcomb E. *Indian Wars.* Boston: Houghton Mifflin Co., 1977.
Utley, Robert M. ed. *Life In Custer's Cavalry, Diaries and Letters of Albert and Jennie Barnitz 1867-1868.* Nebraska: University of Nebraska Press, 1977.

Waldman, Carl. *Atlas of The North American Indian.* New York: Facts On File, 1985.
War Department. *Army Soldier Manual M5 Leadership and the Negro Soldier.* Headquarters Army Service Forces, 1944.
Watkins, Sherrin. *Native American History Reference Manual.* California: Myles Publishing, 1992.
Wharfield, H.B. *Tenth Cavalry and Border Fights.* California: H.B. Wharfield, 1965.
Wheeler, Homer. *The Frontier Trail, Los Angeles.* 1923.
Wilkinson, Frederick D. Ed. *Directory of Graduates, Howard University 1870-1963.* Washington, D.C., 1963.
Williams, George W. *History of the Negro Race in American From 1619 to 1880 V.11.* New York: G.P. Putnam's Sons, 1883.
Woodward, Eleon A. *The Negro In The Military Service of The United States 1639-1886 8 Vols.* Washington, D.C.: Adjutant General's Office National Archives, 1888 (Microcopy T-823).
Worcester, Donald E. *The Apaches Eagles of The Southwest.* Oklahoma: University of Oklahoma Press, 1979.

INDEX

INDEX

A

Adair, Lt., 107-109
Adams, Henry, 115
Alchise, Cochise, 36
Alexander, John H., 116, 260, 371
Allensworth, Allen, 53, 117, 274
Alston, Norman, 119
Alvord, Henry, 29, 30
Amish People, 201
Anderson, Captain, 64
Anderson, George G., 119, 275
Anderson, John, 40
Anderson, Richard, 87, 88
Anderson, William H., 273
Anderson, William T., 119, 276
Apache, Kid, 36
Archer, Sylvester, 120, 121
Armes, G.A., 26, 390
Arnold John, 94
Arthur, Charles, 94
Ashe, Richard B., 121
Ashport, Lemuel, 122
Associate Buffalo Soldier, 241, 242, 245, 246
Auto Repair Shop, 356, 357

B

Baker, Charles, 34
Baker, Edward L. Jr., 122, 273
Baker, Newton, 375, 376
Band, Tenth Cavalry, 340, 359
Bartletts' Ranch, 10
Baseball Team, 1902, 328
Baseball Team, 1908, 329
Batchelor, Captain, 102
Battles, Ninth Cavalry, 14,15
Battles, Tenth Cavalry, 45-47
Bell, Dennis, 15, 123
Bentzone, Captain, 60
Berry, George, 40, 123
Berry, George M. Jr., 13
Beyer, Captain, 10, 126
Big Band, 186
Big Tree, 32, 81
Billingslea, Gaines, 123
Billy The Kid, 9, 10
Biographical Sketches, 113-186
Bivins, Horace, 124, 278
Blackburn, George M., 124
Blackburn, Joseph H., 125
Blakeney, Joseph, 125
Black, Jockey, 193
Black, Kettle, 29
Bliss, Z. R., 73, 306
Bogle, Jackson, 125
Boomers, 12
Boone, Daniel, 192
Bowman, John H., 125
Boxer, Rebellion, 103
Boyd, Captain, 42, 107-109
Boyne, Thomas, 126
Brent, Private, 35
Brooks, William P., 126
Brown, Arthur M., 127, 279, 280
Brown, Benjamin, 127, 128
Brown, David, 128
Brown, George, 128
Brown, H. W., 94
Brown, John, 94
Brown, Plummer, 128, 129
Brown, Samuel, 90
Brown, W.C., 42
Brown, Wesley, 145, 309

Browne, Captain, 374
Brownsville Affair, 65
Brownsville, Texas, 65, 66, 162
Bruen, Julia Coleman, 370
Bruton, Alexander, 129
Buck, John, 277
Buckmeier, Philip, 175
Buckner, Benjamin, 129
Buffalo, 81, 243, 323
Buffalo Soldiers Postage
 Stamp 87, 243
Buffalo Soldiers Monument, 243, 244
Bullis, Lt., 52, 181
Bullock, Burwell, 93
Burn, William H., 129
Burton, James W., 129

C

Campbell, J. H., 94
Campbell John P., 131
Canty, John C., 132
Caron, Captain, 43
Carpenter, Charles, 12
Carter, Beverly, 132
Carrizal, 42, 107-111
Carrizal Prisoners, 307
Carroll, Captain, 9, 176
Carter, Beverly, 132
Carter, Isaiah, 132
Carter, Louis A., 132, 133, 281
Cushwell, Martin, 133
Causby, John, 134
Chaffie, General, 101
Chaplains of Color, 310
Chinn, John, 94
Chowman, 204
Christy, Catherine, 27
Christy, Jacob, 27
Christy, Jacob Jr., 27
Christy, John, 27
Christy, Joseph, 27

Christy, Mary Jane, 27
Christy, Samuel, 27
Christy, William, 26, 27
Churchill, General, 197
Clark, Major, 246, 247
Clark, Powhattan, 37, 38, 262
Clark, Wallace, 116
Cleveland, President, 117, 166, 371
Coalman, Frank, 134
Cochran, Benjamin, 134
Cody, Buffalo Bill, 29
Color Guard, Ninth Cavalry, 312
Color Sergeants, Tenth Cavalry, 322
Conrad, George Jr., 134
Conray, William T., 135
Cook, Eli, 135
Cook, G.W., 111
Cook, William H., 135
Cooper, William A., 94
Couch Captain, 13
Coxey's Army, 39, 63
Coxey, Jacob, 63
Craig, Thomas, 93
Cree Indians, 39
Cress, George, 107
Crippen, Elijiah, 93
Crumbly, Floyd E., 135-136
Cunningham, Charles, 136
Curley, Mayor, 164
Curtis, A.M., 111, 127
Cusack, Patrick, 7
Custer, Elizabeth, 28

D

Daggett, A.S., 95
Dancy, John C., 111
Daniel, Charles, 137, 267
Daniel, J., 266
Davis, Benjamin O. Jr., 202
Davis, Benjamin, O. Sr., 89, 138, 139, 201, 282, 372

Davis, Edward, 93
Davis, Sergeant, 28
Dawson, Captain, 10
Day, James E., 109
Day, Lt., 110
Denigre, A. Walter, 375
Denny, John, 139, 254
Desta Ras, 199
Dodge, F.S., 7
Dodson, John H., 94
Dog Canyon, 11
Dog Soldiers, 80
Dommick, Lt., 88
Dorsey, Charles S., 139
Dorsey, Edward, 140
Douglas, Major, 59
DuBoise, Stephen, 140
Duffin, Elijah, 140
DuVall, Robert L., 140
Dyers, George, 94
Dyers, L.C., 111

E

Eight Cavalry Regiment, 204
Elliston, Amos, 94
Ervine, James W., 93
Escort Duties, 31, 60, 85
Evans, Lt., 33

F

Factor, Pompey, 141, 181
Fall, Albert, 144, 388, 389
Fasit, Benjamin, 40
Field Mess, 358-360
Fern, Henry, 94
Finley, Lt., 38
Fitz, Lee, 141
Fleming, R.J., 94
Flipper, Henry O., 142-145, 253, 261, 389, 390
Ford, George W., 146, 147
Forest Fires, 43, 44, 67
Forsyth, George A., 29
Fort, Lewis, 93
Fort Myer, Machine Gun Troop, 330
Foster, Saint, 95, 283
Frances, Captain, 88
Franklin, Benjamin, 94
Froehlke, Robert, 65
Fuller, John R., 147

G

Gaither, O.G., 40
Gandy, James, 93
Garrison, Lindley, M., 41, 164
Gaskins, B.F., 94
Geronimo, 36, 37, 81, 85, 178
Gettysburg, Pa., 201
Givens, Gilmore, 94
Givens, William H., 95
Gleeton, Walton, 109, 110
Gomez, General, 107
Gould, Luther, 94
Graham, Captain, 29
Graves, Nathan, 147
Greaves, Clinton, 147
Green, Cecelia, 271
Green, John E. 366
Gregory, William, 94
Grierson Benjamin, 26, 36, 38, 52, 61
Griffin, James C., 299
Gross, Philip, 148
Gross, Robert G., 148
Guilfolge, John, 12, 382, 383, 385
Gunter, U.G., 94

H

Haiti, 373
Hammond, Wade, 148, 284
Hampton, Institute, 366
Hampton Institute Indian Students, 270, 366
Hardy, Thomas, 94
Harrington, William, 239, 243
Harris, Richard, 148
Harris, Sherman, 94
Harris, Willie E., 149
Harshfield, John, 31
Hart, Alphonso, 370
Hastie, William, 366
Hatch, Edward, 7, 11, 13, 88
Hatcher, Willis, 94
Hawkins, Emment, 53
Hawkins, John R., 110
Hayes, Arthur, 149
Hayes, R. B., 12
Heller, Joseph M., 102
Henry, Boston, 8
Henry, Guy, 186
Heraldie Items, 5, 26
Herbert, Thomas, 40
Herman, Ltc., 43
Hickok, Wild Bill, 30, 86
Hill, Genevieve, 191, 205, 206
Hill, J. H., 286
Hines, Will, 109, 110
Hisher, Wiley, 94
Hodge, Samuel, 150
Hopkins, Charles, 94
Horse, John Chief, 71
Horse-Shoeing, 355
Houston, Adam, 40
Houston Riots, 54, 55
Howard, Oliver O., 367
Hubbard, Governor, 8
Huguley, John W. III, 258
Hungerford, Captain, 43

Hunt, Tenth Cavalry, 331

I

Ignacio, Chief, 9
Immigration, 197
Indian Wars, 83-90
Irvin, Hoyle, 93

J

Jackson, Emmett, 150
Jackson, John, 150
Jackson, Lewis, 150
Jackson, Lon, 151
James, John, 67
James, Lt., 55
James, Richard, 94
Jefferson, Charles, 93
Jefferson, John, 269
Jeter, John A. 42
Johnson, Andrew W., 151
Johnson Congressman, 110
Johnson, Chaarles S. Jr., 136
Johnson, County, Wyoming War, 366
Johnson, Elias W., 151, 152
Johnson, Gilbert, 366
Johnson, Henry, 151, 256
Johnson, James, 93
Johnson, Joseph, H., 152, 295
Johnson, Nathan, 6
Johnson, Private, 287
Johnson, Smith, 94
Johnston, Mose, 152
Jones, Allen, 94
Jones, Elsie, 40, 94
Jones, J. H., 95
Jones, Lt., 34
Jones, Walter C., 152
Jones, Wesley, 94

Jordan, George, 153
July, Billy, 265
Jumper, John, 72

K

Kassenbaum, Nancy, 244
Kelly, Daniel, 153
Kennard, John, 374, 375
Kennedy, John, 115
Kicking Bird, 31
Kidd, Meredith, 31
Kimble, Lt., 8
King, William B., 199

L

Landon, Isaac, 153
Lane, Ed. 94
Langston, John M., 166
Lawton, General, 103
Lebo, Captain, 34, 37
Lee, Captain, 34
Lee, Robert E., 94
Lewis, Frank, 154
Lewis, Lt., 67
Lewis, Sprague, 94
Liberia, 374, 376
Light Machine Gun Troop, 352, 353
Lincoln County, 9, 10
Logan, Charles, 154
Logan, John A., 154
Logan, Rayford, 66, 67
Lone Wolf, 32
Lord, Captain, 186
Love, Frank W., 155
Loving, Walter, 155, 288
Lucas, Gurnzee, 155
Lyon, Lt., 53

M

Macer, Wilma, 272
Mack, Manuel, 156
Mackenzie, Colonel, 52, 60, 72
Madden, martin B., 102
Maiden Head, 204
Mangas, 36, 37, 81
Mapes, W. S. Lt., 67
Marshall, Lewis, 94
Marshall, Thurgood, 115
Mason, John, 93
Mason, Patrick Jr., 156
Matthews, Charles, 109, 110
Matthews, Mark, 189-207
Mays, Isaiah, 157, 255
McArmack, Henry, 94
McBride, Thomas, 157
McBride, Tom, 158
McBryar, William, 158, 289
McCan, Peter, 40
McDowell, Edward A., 290
McDowell, Martin, 159
McElroy, Hugh, 159
McLain, William, 160
McSween, Alexander, 9
Menu, 79
Merritt, Colonel, 7
Miholland, John, 65
Milbrown, Robert, 94
Miles, Nelson General, 186
Miller, Thomas, S.P., 291
Milliman, Jerry, 160, 161
Minor, Samuel, 94
Mitchell, J.G., 94
Mizner, J. K., 38
Moore, Joseph M., 292
Moore, Miles 161, 162, 263
Morey, L. S., 107
Morgan, Jones, 243, 244
Morrow, Major, 8, 11
Mosby, Corporal, 378

Mounted Drill, 341
Mountain Sheep, 383, 384
Murray, Daniel, 102

N

Nana, 12, 81, 87, 131
NCO Staff and Band, 319
NCO's Tenth Cavalry, 327
Nelson, Edward, 93
Nelson, Marcus, 206
Nelson, Sergeant, 198
Ninth Cavalry Regiment, 313, 13-21, 101
Nolan, Captain, 34
Northern Pacific Railway, 39, 62, 63

O

Olsen, Herder, 383
Oregin of a name, 87
Original Buffalo Soldiers, 241

P

Pacer, Horse, 193
Paine, Adam, 162
Parker, Captain, 40, 172
Parker, R.A., 12
Pass In Review, 338, 339
Payne, David, 12
Payne, Isaac, 163, 181
Payne, William, 40
Pendergrass, John C., 293
Peppin, Sheriff, 10
Perea, Beverly, 111, 163, 164
Perry, Captain, 71
Pershing, John, 39, 109, 308, 386, 387
Philippine Insurrection, 101-103
Pierce, John, 164

Plains Indian, 75-81
Plummer, Henry, 164-166, 294
Pogue, Peter, 166
Polk, Edward, 167
Polo Team, Tenth Cavalry, 332
Powell, Colin, General, 244-246
Preston, Emmett, 167
Pridgen, John T., 366
Prince, Noah, 93
Prioleau, George W., 167, 295
Proctor, John C., 296
Punitive Expedition, Mexico, 41, 105-111, 368
Purington, George, 9

Q

Queen, Howard D., 168
Queen, Mary, 199

R

Rainey, William, 95
Rankin, Frank, 94
Ray, John T., 168
Red River War, 162
Reed, Ernest, 169
Reese, Major, 390
Reid, George, 169
Remington, Frederic, 368, 369
Remount Training, 347, 348
Reynolds, Robert, 169
Rhea, Robert, 169
Rhoten, Marian C., 170, 289
Riddell, Houston, 94
Ridgely, Frank, 94
Rivas, Ltc., 107
Robb, Philip, 170
Roberts, William, 109, 110
Robinson, Aaron, 170

Robinson, Benjamin, 170
Robinson, Charles, 94
Robinson, John, 94
Roman, Nose, 81
Roosevelt, Franklin, 205
Rooevelt, Theodore, 65, 66, 193
Rucker, Dewitt, 8, 109

S

Sam Crow, 39
Sanders, John C., 171, 297
Satank, 32, 81
Satanto, 31
Saunderson, Peter, 94
Sayles, Robert, 171
Scarborough, William S., 116
Schofield, Major, 33
Scott, Charles, 171
Scott, Hugh L., 111
Scott, Oscar J. W., 171, 283
Scott, Winfield, 37
Segura, Cayetarra, 35
Selassie, Haile, Emperor, 199
Seminole Negro Indian Scouts, 69-73, 85, 245, 246, 268
Sequoia and Grant National Parks, 372, 377-381
Seventh Cavalry, 28, 204, 371
Shacker, Fred, 94
Shafter, Colonel, 8, 25, 142
Shaw, Thomas, 172, 173, 257
Shears, James, 173
Sheridan, General, 30
Sherman, General, 32
Shoulder Blade, 39
Simms, Edward, 173, 174
Simms, John, 174
Sinclair, Thomas, 93
Single, Footer, 193
Slaughter, William H., 94
Smith, A J., 298
Smith, Columbus, 174
Smith, Corporal, 378
Smith, George Lt., 9
Smith, George H., 174
Smith, Hosea, B., 44
Smith, Jacob C., 299
Smith, Lt., 88
Smith, Private, 40
Smith, Richard, 174
Smithers, R. G., 34
Smoot, John, H., 94
Snowden, Howard, 175
Sowawick, Chief, 382
Spankler, Robert, 175
Spanish American War, 91-97
Spanish Blockhouse, 311
Spear, James, 93
Spence, Percy, 175
Stage Coach Escorts, 6
Staff Band, Tenth Cavalry, 326
Stance, Emmanuel, 175, 176
Stanley, D. H., 73
Stevens, Jacob W., 176
Steward, Chaplain, 88, 94, 253
Steward, George W., 378
Stroal, George, 94
Sturgis, Harry, 94
Summer, John, 176
Swain, D. G., 390
Swann, M. R., 200

T

Tabby, Chief, 382
Tancil, George, 177
Taylor, Isom, 94
Taylor, John L., 94
Taylor John T., 94
Tenth Cavalry Hospital Corps, 320
Tenth Cavalry Regiment, 23-48, 342-345, 351, 361
Terraza, Juaquin, 36

Terrell, Jack, 193, 194
Terrell, Robert, 111
Tiell, Jacob, 93
They Did Not Tell Me, 365
Thirteenth U.S. Cavalry, 368
Thomas, Alfred, J., 41
Thomas, Harvey A., 177
Thomas, William M., 177
Thornton, George B., 177, 178
Thurmont, Maryland, 201
Tokyo Rose, 203
Toliver, William, 178
Tompkins, Major, 42, 368
Tompkins, William H., 264
Transitional Buffalo Soldier, 241
Trophies and Awards, 324
Troopers, Tenth Cavalry, 346, 349-350, 354, 362
Troutman, Edward, 7
Trumpeters, Tenth U.S. Cavalry, 325
Tucker, Willie, 12
Turner, Charles B., 178, 300
Turner, Henry M., 165, 166
Turner, William H., 93
Twenty-fourth Infantry Regiment, 49-55, 101, 320
Twenty-fifth Infantry Regiment, 57-68, 101, 314-318
Tynes, James F., 179
Tyler, Charles, 179

U

Uncompahgres, 382
Uintahs, 381, 382
Union Pacific Railroad, 29
Utah Frontier, 381-386
Utes, Colorado, 8, 9
Ute Indians, 40, 41, 152

V

Van Vliet, Major, 37
Victoria, 11, 26, 52, 61, 81, 85, 86, 126, 131, 142, 178
Villa Poncho, 42, 193, 203, 204
Vrooman, William A., 179

W

Waites, James, 180
Wadsworth, Senator, 388
Walley, Augustus, 180
Walker, Corporal, 40
Walker, Sgt., 94
Wanton, George H., 180
Ward, John, 181
Warfield, Edward, 66
Warmsley, William C., 182, 325
Warren, George, 93
Washington, Bob, 182
Washington, E. S., 94
Washington, William, 301
Watson, John, 94
Watson, mary, 191
Weaker, Corporal, 35
Weaver, Agent, 8, 9
Weaver, Asa, 12
Weeks, John W., 388, 389
West, Benjamin, 94, 183
Whirlwind 39
White Settlers 8, 12, 13, 31, 61, 62, 85, 175
Whitson, J. T., 370
Wilcox, William C., 302
Wiley, John T., 182
Williams, Arthur, 183
Williams, Bill, 266
Williams, George W., 183-184, 274
Williams, Isaiah, 184
Williams, John W., 88

Williams, Joseph, 94, 184
Williams, Sharpe, 375, 376
Williams, Walter, 303
Willis, Dorsie, 66, 321
Wilson, Alfred, 93
Wilson, William O., 13, 186, 187
Wilson, Woodrow, 375, 376
Woods, Brent, 259
Woodard, Lt., 32
Wounded Knee, 13, 86, 179, 186
Wright, Emanuel, 6
Wright, Marcellas, 94
Wyatt, Nathan, 94

Y

Yellow Hair, 39
Young, Ada, 386, 387
Young, Charles 42, 101, 102, 304-306, 368, 370-387
Young, Gabriel, 370
Young, Peter, 370
Young, Susan, 370